SEVEN DECADES OF MILK

A HISTORY OF NEW YORK'S DAIRY INDUSTRY

JOHN J. DILLON

Editor The Rural New-Yorker

1941

Copyright © 2013 Read Books Ltd.
This book is copyright and may not be
reproduced or copied in any way without
the express permission of the publisher in writing

British Library Cataloguing-in-Publication Data
A catalogue record for this book is available from the
British Library

Cattle Farming

Cattle are the most common type of large domesticated ungulates (a group of mammals, mainly categorised by their hoofs). They are a prominent modern member of the subfamily *Bovinae*, and are the most widespread species of the genus *Bos*. Cows, or cattle, are commonly raised as livestock for meat, as well as dairy animals and even draft animals, kept for such errands as pulling carts, plows and the like. Other products include leather and dung for manure or fuel. From as few as 80 progenitors domesticated in southeast Turkey and northern Iraq about 10,500 years ago, an estimated 1.3 billion cattle are in the world today. Cattle occupy a unique role in human history, they are one of the few animals to have been domesticated since at least the early Neolithic period – and they have been seen variously as workers, sacred animals, and foodstuffs.

Cattle are often raised by allowing herds to graze on the grasses of large tracts of rangeland. The most common interactions with cattle involve daily feeding, cleaning and milking. Many routine husbandry practices involve ear tagging, dehorning, loading, medical operations, vaccinations and hoof care, as well as training for agricultural shows and preparations. Interestingly, there are many cultural differences which occur when working with cattle; the cattle husbandry of Fulani men rests on behavioural techniques, whereas in Europe, cattle are controlled primarily by physical means, such as fences. In terms of food intake by humans, consumption of cattle is less efficient than of grain or vegetables with regard to land use, and hence cattle grazing consumes a larger area than such other agricultural production, especially when they are raised on grains. However, cattle and other forms of domesticated animals can sometimes help to use plant resources in areas not easily amenable to other forms of agriculture.

Cattle today are the basis of a multi-billion dollar industry worldwide. The international trade in beef for 2000 was over $30 billion and represented only 23% of world beef production. The production of milk, which is also made into cheese, butter, yogurt, and other dairy products, is comparable in economic size to beef production, and provides an important part of the food supply for many of the world's people. There are some pressing trepidations concerning cattle farming though. A report from the 'Food and Agriculture Organization' (FAO) states that the livestock sector is 'responsible for 18% of greenhouse gas emissions', and the report concludes, that unless changes are made, the damage may more than double by 2050, as demand for meat increases. Another concern is manure, which if not well-managed, can lead to adverse environmental consequences. These are issues which both farmers and governments are working on though, so that cattle farming and the commercial usage of cows can, from its long and largely inimitable roots, progress and develop into the future.

I have written this book in appreciation of my seventy years' association with the milk producers of New York State, and I am, therefore, pleased and honored to dedicate it to those producers and their families

The author gratefully acknowledges the generous and valuable help of William F. Berghold, Miss M. Gertrude Keyes, and Mrs. E. T. Royle in checking records, preparing copy and reading proofs. Acknowledgment is also due to hosts of dairy friends whose suggestions, counsel and information have been helpful and inspiring.

CONTENTS

CHAPTER		PAGE
	FOREWORD	xi
I.	THE FIRST MILK RECORDS	1
II.	THE FIRST MILK ORGANIZATIONS	5
III.	LAWS, REGULATIONS AND INVENTIONS	19
IV.	THE GROWTH OF INDUSTRY	28
V.	THE O'MALLEY INVESTIGATION	33
VI.	MILK PRICES AND GRADES	40
VII.	MILK FREIGHT RATES	51
VIII.	STUDY OF DISTRIBUTION ADVOCATED	55
IX.	THE DEPARTMENT OF MARKETS	62
X.	MILK CAMPAIGN STARTED	67
XI.	THE DAIRYMEN'S LEAGUE, INC. IN 1915	75
XII.	THE MILK FIGHT OF 1916	84
XIII.	ORGANIZED FARM CO-OPERATION	104
XIV.	UNITED DAIRYMEN HOLD A MEETING	108
XV.	FARM LEADERSHIP WEAK	112
XVI.	POLITICS RUN RIOT	119
XVII.	THE FEDERAL MILK COMMITTEE	132
XVIII.	THE COUNTRY MILK COMPANY	140
XIX.	THE 1919 MILK STRIKE	150
XX.	ADMIT FAILURE: SEEK NEW POWER	159
XXI.	THE MAJOR MILK TRAGEDY	166
XXII.	POOLING AND CLASSIFICATION BEGIN	178
XXIII.	THE BORDEN-LEAGUE ALLIANCE	186
XXIV.	SHEFFIELD FARMS STOOD ALOOF	195

CONTENTS

CHAPTER		PAGE
XXV.	Laying the Basis of Monopoly	200
XXVI.	State Milk Control	209
XXVII.	Governor Vetoes Farm Bill	217
XXVIII.	The Rogers-Allen Law	225
XXIX.	Bargaining Agencies Collapse	234
XXX.	The Federal-State Orders	240
XXXI.	The Legal Entanglements	250
XXXII.	Desperate Farmers Rebel	264
XXXIII.	Three Official Reports	276
XXXIV.	Dairy Laws and the Courts	300
XXXV.	Dealers' Schemes	311
XXXVI.	Essential Principles Violated	323
XXXVII.	God Helps Those Who Help Themselves	326

FOREWORD

Many dairymen, widely distributed, have urged me for several years by letter and speech to write a history of the milk industry. To comply with their requests I have been obliged to do the work at such irregular times as could be spared from my editorial and publishing duties and other interests. While my memory covers seven decades, I have been obliged to rely mainly on disconnected records and boyhood memories for the information up to 1870. Since that date my contact with both production and distribution in the industry has been practically uninterrupted. From this experience I would prefer to forget the trials, the hardships and the injustices resulting from the distribution of milk, and confine myself to a story of the cheerful homes, the beautiful lives, and the noble men and women it has been my good fortune to know on the dairy farms of the New York milkshed. But duty, not pleasure, has inspired my task. I rely on facts and truth to work a reform in the distribution of milk, and I have had no choice but to go by the record. I have seen selfishness and greed dominating our dairy industry from its beginning; producers and consumers of milk have been continually and ruthlessly exploited. These abuses can be corrected without hardship to anyone or any class and with justice and profit to all.

Milk is one of our most important foods. It is an essential in the home of infants and children and an important element in the daily diet of every member of the family.

The New York milkshed is especially well adapted to its production. Our farmers are skilled and successful in its production. The labor is continuous and hard. The control of distribution and prices is a rightful function of producers, but producers are widely distributed, so organized dealers trading with individual farmers gained control of the instruments of distribution. This gave dealers power to usurp the producers' inherent rights and to fix the price farmers receive and the price consumers pay. This is double monopoly.

Therein is the crux of the milk problem. Milk dealers conspire to fix the price they pay farmers for milk. Farmers contend for their right to set the price on the milk they produce.

No one will dispute that under our institution of private property the producer is entitled to the fruits of his skill and labor. So any act that denies the producer of wealth that right is clearly a violation of the Constitution. Formerly when the question was raised the State and Federal Courts upheld these basic laws. In 1916, united farmers demanded a price set by themselves for their milk, fought for it and won. Later, however, a combination of twenty-four self-appointed farm leaders made an alliance with dealers. The combination organized a counterfeit farm co-operative corporation. Through deceit, intrigue and misrepresentations it induced some farmers to adopt this corporation as their agents for the sale of their milk. The Borden Company and some others forced their producers to do the same. The details of this bogus organization is related in the text. It is enough to say here that this organization is now the basis of a chain of corporations which with the State and Federal Government constitute our present milk monopoly in which basic prices of milk and milk products are fixed on bogus exchanges at Plymouth and Chicago. No producer in the state is permitted to sell a quart of milk in the New York market except through this monopoly and at prices and terms fixed by it.

It seems to me that this is centralization in the extreme. If it should be applied to our markets generally, it would destroy our whole American Economy. It would be but one step to totalitarian civil government—to communism. I believe that a democracy cannot long endure with a communistic control of its markets and its industries.

JOHN J. DILLON.

CHAPTER I

THE FIRST MILK RECORDS

One Hundred Years Ago.

One hundred years ago cows were tied to stakes in the City of New York and fed on garbage. Owners of property on Christopher Street rented the privilege of herding cows in the street. The disposition of the manure was a provision of the lease contract. At that time there was a demand for the manure on the farms cultivated on the land now covered by St. Patrick's Cathedral on Fifth Avenue and 50th Street, and the Empire State Building at Fifth Avenue and 34th Street.

Historically, however, the milk business in the City of New York goes back to the primitive days when people milked their own cows or bought milk from their next door neighbors. This practice is roughly fixed as late as 1750 to 1800. In 1806 we read that deliveries were so small that for the most part milk was carried by hand. As the demand increased distributors used a wooden yoke. This was a piece of wood about three feet long chiselled out and smoothed to fit over the shoulders and the back of the neck, with nicely rounded arms extending over the shoulders. A light chain or a rope was suspended from each arm with a hook at the end. With this yoke across his shoulders the carrier stood between two pails or other containers, and by stooping forward attached the hooks to the vessels; then straightening up, the weight of the vessels rested on his shoulders. He steadied the load by his hands. In this way he carried several gallons at a time. I do not know whether or not this method of delivering milk was ever in use in other parts of the country, but I distinctly remember seeing this type of yoke used seventy years ago in Sullivan County by Milton Gillespie, a neighboring farmer, for carrying slops to the pigs and other uses about the farm. It is interesting to note that on December 23, 1763, the Common Council of Manhattan fixed the price of milk at six English coppers a quart, or about 12 cents of our money.

Manhattan Island was not a good grazing territory. As the population of the city on the southern point of the island increased, the farmers were pushed back. There were no means of transporting milk from the country. Dairymen were forced to feed their cows on such materials as were available. Finally they discovered that cows would eat brewers' grains,—thereafter known as "slop feeds,"—and produce more milk than on other available feeds. This food was abundant. During the 1830's there were 18,000 cows in New York City and Brooklyn fed on this food almost exclusively. The cows were crowded into crude stables erected for the purpose, but once in their narrow stalls never left them alive. The large stables were built near the breweries and in some cases the sloppy food ran from the breweries to the cows' mangers through wooden chutes.

By 1842 the Erie Railroad had been extended as far as Goshen, Orange County. It was a tradition in Orange County sixty years ago that the first shipment of milk from that county to New York City was made in a churn in the year 1842. The milk was not popular. The consumers complained that a yellow scum gathered on the top of it, when held for a time. Cows fed on brewery waste did not produce milk rich in butter fat.

In 1844 the Orange County Milk Association was organized with a capital of $5,000 to ship milk to New York from Orange County farmers. Owing to lack of proper train service by the Erie Railroad and the inexperience of the farmers themselves, the association lost money during the first years of its operation. The association persevered, however, and later became successful and prosperous.

About the same time, considerable quantities of milk were shipped into the city by farmers of the Harlem River Valley over the New York and Harlem Railroad. In 1847 these shipments averaged about 50,000 quarts daily.

Before it began to ship milk in large quantities, Orange County was famous for the volume and quality of its butter.

THE FIRST MILK RECORDS

In 1876, when I first began to cart milk from the Reverend Augustus Seward's Round Hill Farm, near the village of Florida, to the Decker Creamery in that village, milk cans had taken the place of churns. The 40-quart cans were of the present type. A full can went on record as 40 quarts; but less than full cans were measured by plunging a gauged flat stick to the bottom of the can. The figure at the moisture point of the stick showed the quarts of milk in the can. Weighing milk as now was not practiced.

The creamery floor space consisted of a vat about the depth of a milk can. This was filled with water nearly to the level of the floor. The milk was passed from the farmer's cans into smaller light-weight tin cans provided with a wire-bale handle so the milk could be lowered into the vat, and removed again for shipment after cooling. At that time the milk was transported to the city on night trains in the 40-quart bare cans. Soon afterwards felt jackets were provided to preserve the low temperature of the milk.

Farmers had platforms erected at crossroads and other points. The milk was delivered on these platforms and the milk trains stopped and loaded them on the cars. The empty cans were supposed to be returned to the place of shipment, but accidents, carelessness, and other uses of the cans in the city caused farmers much annoyance and loss. To correct these abuses farmers induced the Legislature to pass the Milk Can Law which forbade milk cans properly branded to be used by any other person than the owner or for any other purpose. Later on, under new conditions this law was unfairly used to annoy farmers. The shipment of milk in iced cars was a later development. Shipment by large insulated tanks is a still more recent practice.

A Period of Milk Prosperity.

For a time before and following the Civil War the sale of fluid milk was profitable. Where farmers could conveniently deliver milk to the consumer, they bargained with the housewife for the price, did their own collecting and went home with one hundred per cent of the money in their pockets. The farmer dipped the milk out of the can with a long-handled dipper and poured it into the consumer's vessel. Often the pail or pitcher was left outside the door so the delivery could be made in the early morning before the family's hour for rising. In some cases the money to pay for the milk was left in the pail which was to receive the milk.

When the Orange County and Harlem Valley shipments of milk began in the early 1840's, farmers negotiated the price and terms of sale to city distributors. For twenty-five to thirty years the business was profitable to both producer and dealer. The consumer soon learned to like the taste and flavor of fresh, rich, wholesome milk direct from the farm. It came largely by rail, but for a considerable time milk was shipped by the steamship Mary Powell from Newburgh and other Hudson River points. Deliveries were made by horse and wagon in substantially the same way that farmers delivered to consumers in the smaller, nearby villages and in the up-state cities.

As the population of the city grew, the demand for milk increased. The business of distributing milk became more profitable. Dealers multiplied in numbers. Competition increased and the ambitious became eager for larger profits. Rumor had it that distributors diluted the milk by skimming cream from the top of the cans before delivering to consumers. There were complaints that skim milk had been mixed with the whole milk as it came from the cow, and the mixture sold to consumers as whole milk.

CHAPTER II

THE FIRST MILK ORGANIZATIONS

Thirty Vital Years.

Up to 1870 the prices for milk shipped to the New York market were negotiated by farmers as sellers and dealers as buyers. The prices were fair, and the farmers were prosperous. During the early 70's, however, the dealers began to press for lower prices. Farmers resisted.

Amzi Howell, a unique figure in the trade, got up what he called a "Joint Price Committee" for the purpose of stabilizing prices. It was to include farmers and dealers. The committee was practically Mr. Howell. Occasionally consulting a few of his friends in the trade, he announced a price which met the approval of the dealers. Farmers had no part in it. The resistance of producers, however, had a restraining influence, so while reductions had been made the price was satisfactory up to about 1870.

The last three decades of the nineteenth century revolutionized the New York milk business. During these thirty years the dealers acquired full domination of the industry. They were fertile in resources to increase and maintain their mastery of the business when once gained. They gave the farmer no concession whatever. They refused even to sit in with groups of farmers to discuss prices. At rare times they would attend a meeting. To the alternate pleadings and demands of producers for a voice in the councils that fixed prices, they replied that there was nothing to discuss.

"Prices of milk," they said, "are fixed by supply and demand, and no amount of consultation will change it."

They were alert, aggressive and enterprising in their own behalf. Many of them were good fellows, pleasant companions, liberal in charity and in public affairs, generous in social matters and in politics. There were many buyers and distributors who made a practice of buying milk on contract. They paid promptly for a time to win confidence. Later, having won the longest possible amount of credit, they faded out of sight with accumulated milk bills unpaid, or went through bankruptcy. There were in the trade at that period, and are at the present time, many men with scrupulous regard for their credit. But they never tired of effort or scrupled to buy a dollar's worth of milk for 64 cents.

The means and methods which they used to promote their objective were direct and crude. It is still easy to identify their fundamentals in the now refined and indirect methods of their present-day successors. One and all of the big fry as units have pushed their advantage in purchasing power from time to time until farmers have been driven to open revolt, and have won the sympathy of the public. Then they make temporary concessions only to repeat their previous aggressions.

The one obsession of milk dealers for seventy years has been as low a price as they can force upon the producer in the country, and as high a price as they can wring from the consumer in the city.

The Orange County Milk War.

During the late 70's and the early 80's the main portion of the New York City milk supply came from Orange County. When the farmers first abandoned their profitable butter production and began to ship milk, prices were favorable for a time, and dairy farmers prospered; but after the organization of the Milk Exchange, Ltd. prices fell to such a low level that the pent-up resentment of producers broke out in open rebellion. Deliveries stopped. Some producers resisted persuasion. Farmers started with their milk loads for the depot but never got there. Their milk flooded the gutters in the streets of Campbell Hall, Goshen, and other places. The city was short of milk. Farmers told the dealers to pay the price or they would bring their horses and wagons to the city the next day and deliver milk to the consumers. The dealers paid the price.

THE FIRST MILK ORGANIZATIONS

Then the dealers induced the railroads to run milk trains from up-state counties and carry milk at the cost per can that Orange County farmers were paying. Farmers of Sullivan, Delaware, Chenango, Otsego, and other counties were offered higher prices than their milk was worth for making butter and cheese. They yielded to the appeal, burned their churns and bought milk cans. With this supply dealers had a surplus of milk for fluid consumption. The price dropped lower than ever before. Farmers felt they had been tricked and misled. They resented the treatment and were fighting mad.

The Milk Exchange, Ltd.

The Milk Exchange, Ltd. was organized March 11, 1882. It was the first incorporated distributors' organization. The incorporators and original subscribers to the capital stock were: George Slaughter, John W. Tayntor, George Conklin, Charles H. C. Beakes, Robert F. Stevens, Thomas O. Smith, R. R. Tone, W. A. Wright, P. E. Sanford, J. D. Miller, Joseph Laemmle, G. O. Olmstead, T. J. Tuthill and Jesse Durland. The original pretense was that the Exchange was to consist of milk dealers and milk producers, but when it came to writing by-laws the dealers refused to give the farmers one-half the board of directors. Dairymen, therefore, saw the purpose and refused to put themselves in a position to authorize dealers to fix the price to be paid them for milk.

According to the charter, the purpose of the corporation was, "the buying and selling of milk at wholesale and retail, the purchase of dairies of milk when deemed advisable and the sale of same to milk dealers."

The by-laws of the Milk Exchange, Ltd. authorized the board of directors to "fix the market price at which milk shall be purchased by the stockholders" and provided that stockholders purchasing milk at any other price would forfeit their stock and membership. After an investigation by a Senate committee this by-law was revised as to the forfeiture of stock, but the authority to fix prices was retained. Prices so fixed continued to be the prevailing price in New York.

In January, 1891, Attorney General Charles F. Tabor brought suit in the name of the people of New York State to annul the charter of the Milk Exchange, Ltd., charging it to be an "unlawful and illegal combination and conspiracy made in restraint of trade to limit the supply of milk and to fix and control the price thereof in the city of New York and elsewhere."

Finally, on May 1, 1895, the Supreme Court in Broome County entered a decree of dissolution. The certificate of dissolution was filed in the office of the Secretary of State on May 22, 1895. The Milk Exchange, Ltd. was out of existence.

During January, 1892, farmers of northern Pennsylvania, producing milk for the Philadelphia market, protested against the price being paid by the Philadelphia dealers. During the fight that followed, the New York Milk Exchange, Ltd. arranged with its members to ship milk to the Philadelphia market. The shipments were between 200 and 300 cans daily. The cost of this milk laid down in Philadelphia was more than the farmers were asking, but it defeated the regular producers for the market.

After the Exchange had defeated the Pennsylvania farmers with New York milk produced by New York farmers, the Exchange reduced the price to New York producers in mid-winter from three and one-half cents to three cents a quart. New York farmers being under contract with the dealers at the time, were helpless to prevent the Exchange from using their milk to frustrate the efforts of the Pennsylvania farmers to obtain a fair price from the Philadelphia dealers.

In 1887, I rode all night in a day coach of the New York, Ontario and Western Railroad from Middletown to a protest dairy meeting in Delhi, Delaware County. It was a spontaneous assembly of rugged, sterling, and determined farmers who felt that they were the victims of an intrigue and said so.

Out of that meeting and others that followed, the Milk Producers' Union was organized. Northern New Jersey and the Harlem Valley joined the movement. The farmers lacked organization experience. Representatives, stooges of the dealers, worked themselves into the farmer's confidence, meetings, and committees. Some of them were residents of the county. Suspicions of their purposes tended to retard progress, and when the time came to strike, all were not ready. The guns had been spiked. The movement failed.

The Consolidated Milk Exchange, Ltd.

The Consolidated Milk Exchange, Ltd. was organized under the laws of New Jersey on November 15, 1895. This was about six months after the ouster of the Milk Exchange, Ltd. It was evidently organized by the same interests and for the same purpose. The capital stock was $25,000. The principal place of business was Jersey City, N. J. The names of the incorporators were: John A. McBride, J. E. Wells, Thomas B. Harbison, Charles H. C. Beakes, William C. A. Witt, M. L. Sanford, J. V. Jordan, Fred H. Beach, John P. Wierck, George A. Slaughter, and William A. Wright. Four of these were incorporators or members of the ousted corporation. Alfred Ely was attorney for both corporations.

The charter provided that the principal part of the business was to be conducted at Jersey City. It was also authorized to do business in the cities of New York and Brooklyn, in New York, and in all States and foreign countries. The purpose, according to the charter, was to foster and promote trade and commerce in dairy products; to deal in milk and dairy products; to act as commission merchants for the sale of milk and dairy products; and also as agents for farmers, producers and shippers.

Thus ostensibly organized to buy and sell milk, to do a commission business and promote trade in dairy products, it never bought or sold a can of milk. It had a small room at No. 6 Harrison Street in the market district in New York City where the members met from time to time on call, and fixed the price of milk to producers. No milk was ever sold there. Its activities were more fully developed during the early part of the following century. The part played by the Consolidated Milk Exchange, Ltd. in the milk industry during the early part of the twentieth century will appear in following chapters.

At times, when because of drouth, low prices, or other reasons the supply of milk became short, a meeting would be called and a very substantial increase would be made on the price of milk to producers. This would be widely published. Cattle dealers would then be sent into Pennsylvania and other distant points to bring droves of milkers into the milk-producing shed.

Because of the high price offered by the "Exchange," farmers were induced to stock up with these cows, to put in carloads of brewers' grains and commercial feeds. The result would be a large increase in the volume of production. Then another meeting of the Exchange would be called and the price drastically reduced. This procedure was repeated many times. I do not recall that it ever failed to work.

Another stunt that produced desired results, that is to say, low milk prices to farmers, worked out in this way: In the early spring dealers would go into the producing sections and advertise a day to sign contracts for milk, advising farmers where they would sit. The price for these contracts was based on the "Exchange" price, usually less but occasionally a trifle more. Following the day of contract the dealer would visit the big producers who had neglected to call on the dealer. Then the negotiations would run like this:

"Well John, I came to buy your milk."

John would say, "I think I'll go back to making butter and raising calves and pigs. There is no profit in shipping milk; your price is too low."

The dealer would say, "John, you know the price is made by the Exchange. I cannot control that, but you have a good dairy and I will pay you a premium of a quarter of a cent a quart above the Exchange price for three months and an eighth of a cent for four months, and the Exchange price for the five months when surplus is highest."

He usually got his contract. When all of the dealers had covered their supply territory with their contracts to pay Exchange prices, a meeting of the Exchange would be called at No. 6 Harrison Street, and in ten minutes or less the Exchange price would be fixed until further notice.

Of course, the farmer had signed a contract to sell his milk for the year at any price the dealers collectively chose to pay. It was

THE FIRST MILK ORGANIZATIONS

not unlike the classified price plan, where the farmer is under contract to ship his milk on consignment and to accept whatever the dealer wishes to pay.

Five States Milk Producers' Association.

Now the dealers were stronger than before, and more tyrannical. They extended the milk shed into new fields and began to build milk plants for assembling milk in county districts of five States. They bought milk for 50 cents a 40-quart can, and at times for 39 cents. In 1894 farmers again attempted to regain control of their business. They organized the Five States Milk Producers' Association.

The actual leaders this time were lawyers, professional promoters, and financial brokers, with the whole army of milk dealers in the background. George E. Wells of Goshen, N. Y., a member of the Consolidated Milk Exchange, Ltd. was later admitted to membership after identifying himself and his affiliations with the Exchange. Alfred Ely, of Alfred, N. Y., a large milk producer and attorney for the Consolidated Milk Exchange, was also a member. The field plan was to organize producers shipping by each rail route into a group by itself. In this way there were as many groups as shipping lines.

The farmers were in a humor to protest against the prevailing low prices. They were strong and sturdy in body and mind, and fully determined to fight to a finish. They contributed liberally in money and energy. But they were inexperienced. They listened without suspicion to promoters who had ten-dollar bills in their pockets as inducements to boom the private interests of the dealer who contributed the ten dollars. They had no knowledge of the methods pursued in the promotion of legitimate corporations, or of corporations organized for the purpose of deception and exploitation. They put their full trust in the centralized leaders, but those leaders did not take the producers fully into their confidence. There was something of a mystery about the things that were soon to happen, but were never revealed or realized. Some of the group leaders did complain that one of the promoters was using the time of local officials and funds of the association to further his private business.

In the month of March, 1899, almost five years after the movement had been initiated, the following optional contract was prepared by the organization officials for the signature of producers:

"Know all men by these presents, that I,_____ of the town of _____ and State of_____, have made, constitute and appointed

J. C. Latimer, of Tioga Centre, N. Y.,
Ira L. Snell, of Kenwood, N. Y., and
F. B. Aiken, of Mecklenburg, N. Y.,

my true and lawful attorneys for me and in my name, place and stead to bargain, sell and contract all the milk produced by the cows owned or controlled by me, except the milk I use at my house, for a term not longer than five years, at a price not less than $2\frac{3}{4}$ cents per quart for milk produced during the months of October, November, December, January, February and March, and $1\frac{3}{4}$ cents per quart for milk produced during the months of April, May, June, July, August and September. The milk to be paid for at least monthly on or before the tenth day of each month for milk delivered the month previous. Such milk to be of standard quality and to be delivered in good condition at the milk shipping or receiving station at_____.

"I hereby represent that I own or control _____ cows and I agree that I will deliver all the milk produced by me, except what I consume in my own family. I further agree that I will not increase the number of my cows, for milk production to be sold, beyond 25 per cent of the above number during the existence of the contract to be made by me through my attorneys above named, except on written request of the buyer of said milk; but nothing herein contained shall be construed to prevent increase of cows for manufacture of butter and cheese.

"Giving and granting unto my said attorneys full power and authority to do and perform all and every act and thing whatsoever requisite and necessary to be done in and about the premises as fully to all intents and purposes as I might or could do if personally present, with full power of substitution and revocation, hereby ratifying and confirming all that my said attorneys shall lawfully do or cause to be done by virtue thereof."

These men to whom the power of attorney was granted were the Executive Committee of the organization. It was claimed that such contracts were signed for 22,000 cans of milk daily, which was about four-fifths of the quantity moving to New York City.

A. G. Loomis, Deposit, New York, was president of the association at this time, but the members of the Executive Committee were its most active officials. Mr. Loomis, however, objected to the contract because:

1. The power of attorney was vested in individuals personally,
2. It did not state a high enough price,
3. There was no time limit to the option,
4. The sale might not be executed, but if a farmer sold elsewhere within five years, he ran the risk of being sued for breach of contract.

At a meeting in Binghamton, New York, on October 17, 1899, Mr. Latimer, one of the official leaders, announced that the milk had been sold to the Pure Milk Company. The price was to be three cents a quart for November, December, and January; two-and-one-half cents for September; two and one-quarter cents for April and August; two cents for May and July, and one and three-quarters cents for June; one-quarter cent a quart was to be paid in non-assessable preferred stock of the Pure Milk Company. Shipments were to begin November 15.

Mr. Latimer explained that the company had authorized a capital stock of $30,000,000; $20,000,000 preferred and $10,-000,000 common. He explained also that the financiers had insisted that they get one share of common stock as a bonus for every share of preferred stock they bought at $100 each. There had been no stipulation that farmers would be entitled to that privilege, but he stated that he would see to it that farmers got the same privileges as the bankers. He directed them how to make remittances for stock and urged them to do so. Delegates and producers from many sections of the State were present. Confidence and enthusiasm ran high. Farmers felt that a great success had been made. A considerable time was devoted to the questions as to who was eligible to participate in the sale and who was not entitled to the benefits.

Then a well-dressed gentleman who wore a high silk hat asked me to explain why the agricultural press had not indorsed the proposition as it had developed. I had been working diligently to get the kind of information our *Rural New-Yorker* readers ought to have. I could get all kinds of rumors and intimations of big things coming, but the official leaders were not willing to reveal the information I thought farmers should have. I had discovered, however, that a charter for a Pure Milk Company had been drafted, but not filed, and that it was in a stock broker's office in lower Broadway, New York City, waiting for someone to put up $4,500 cash to pay the filing fee.

I was a younger man at that time but I tried to explain to the gentleman that *The Rural New-Yorker* felt a responsibility to farmers for the information it printed, and that it could not try to influence farmers to do things that we would not do ourselves without exact information, and this we had been unable to get. I said, however, that if exact information could be had, I would be glad to publish full details of the transaction. I asked the following questions and Mr. Latimer replied:

Q. To whom have you sold the milk?
A. The Pure Milk Company.
Q. Who are the officers of the company?
A. Well, the officers have not been elected yet.
Q. Who are the directors?
A. No directors have been elected yet.
Q. Is the charter drafted?
A. Yes.
Q. In what State is the charter filed?
A. It has not been filed yet.
Q. Then as a matter of fact, the Pure Milk Company is not yet in existence.
A. Not legally.

I then explained to the farmers present at the meeting the nature of a corporation, how their option contracts might be used by promoters to get possession of the stock without a cent of money, and the peril to farmers in paying cash for a minority holding in a corporation, leaving the management free to get the majority of the stock for nothing, and control the company and the farmers' milk. I would, I said, advise farmers to organize and invest their money in co-operative associations properly organized

THE FIRST MILK ORGANIZATIONS

and kept under their own control, but not in the minority stock of a milk company controlled by speculators.

My talk was not popular. I got no applause. On my way out I was made to feel that my editorial ethics were not appreciated, but the farmers did approve generally what I said about the stocks, especially my advice to keep their money in their pockets. Of course, November 15 came and went, but no milk was sold. The charter was never filed. Later the men who abused me in their disappointment admitted that my limited information about the proposition was better than theirs, and they made generous apologies for their hasty criticism.

Two years later I met the head city promoter at a winter resort in Lakewood, N. J. He gave me the details. The majority of the stock was to be issued, as I had suspected, in exchange for the contracts. The brokers and the promoters were to have a liberal portion of the stock for service in addition to a regular cash brokerage. The hitch was that no one would put up the $4,500 cash fee for the filing of the charter in the State of Delaware. The last chance was to induce the farmers at the Binghamton meeting to subscribe in advance for enough stock to file the charter.

The Rise of Borden's.

During the last two decades of the nineteenth century the Anglo-Swiss Condensery Company, a wealthy corporation, owned and operated many condenseries in Orange and other counties. One of these factories, located at Middletown, in Orange County, bought milk from farmers within a radius of about ten miles. Local farmers supplied the people of Middletown with milk direct from the farms.

Sometime in the early 90's the Anglo-Swiss Company began to distribute milk in Middletown from the condensery. It made its appeal on the ground that it paid taxes in the city while farmers outside the city limits paid no city taxes. The retail price was five cents a quart. Competition developed, and on May 1, 1899, the company announced that the price would be three cents instead of five cents. The farmers reduced their price to four cents.

Many consumers felt that condensing milk was the business of the company, and that if encouraged to force producers out of delivering milk, consumers would be at the mercy of the company. They continued to buy direct from the farms. Just how long the warfare lasted, I do not recall, but ultimately the Anglo-Swiss Company sold out to the Borden Company and the trouble ended. I record the incident because it was typical of the recurrent efforts of big milk concerns to obtain a monopoly of the milk business by cut-throat competition.

During the last half of the nineteenth century, Gail Borden built and operated condenseries in Orange, Westchester, Putnam and Dutchess Counties. Later the business was incorporated as Borden's Condensed Milk Company. In September of 1899, the company had reduced the prices so low that 80 per cent of the producers east of the Hudson River refused to sign the renewal contracts. In addition to the low price and other rigid regulations the company dictated the feed to be given the cows. It particularly banned brewers' grains and silage. This was one of the producers' grievances.

The fight centered at Brewster, Millerton, and Wassaic. At separate meetings at these places on September 11 of that year, farmers asked the company to raise the price ten cents a 100 pounds over the previous year's price which averaged about three cents per quart. Farmers pleaded that they were suffering from effects of a drouth. Hay was only a half crop and corn was also short. Cows cost $45 to $65 a head, and feed had risen from $13.50 to $17.50 a ton.

The story told at the time by farmers for this small region is much like what might be told now of the whole State. Gail Borden started the business with small financial backing. The farmers were prosperous and well-to-do. They helped Mr. Borden with money and credit. He pleaded with farmers for help, promising future rewards, but he lost control of the business.

The new management reduced prices still lower every six months. An 1800-word contract tied the farmer hand and foot. It bound him to drastic regulations, but bound the company to nothing except to pay for milk that it accepted. It was free to

THE FIRST MILK ORGANIZATIONS 17

refuse any milk it did not want. The reduced price had impoverished the farmers, forced them into debt, and reduced the income and value of the farms. It even reduced their resistance to abuse. They were helpless. The company strongly refused to meet the farmers' demand for a ten cent per cwt. raise, and allowed them a little less than four cents for a three-months period.

Beginning of Price Decline.

In 1889 Dairy Commissioner Brown estimated that there were 1200 dairy plants for shipping and manufacturing milk in the State, and that the annual production per cow was 3034 pounds. The number of cows in the State in 1890 was 1,440,230. The price of milk at the plants during all of this period was fixed by the proprietors and dealers. The price declined for every five year period from 1870 to 1895. During the latter part of the period, however, farmers began to build co-operative plants for the sale of fluid milk and also for making butter and cheese. These reported higher returns to producers than the proprietors' plants and factories regularly over a number of years.

The average prices for commercial milk at local plants per quart were reported as follows:

Year	Cents	Year	Cents
1870	4.60	1885	2.79
1875	3.58	1890	2.63
1880	2.88	1895	2.52

This table was published in *The Rural New-Yorker* before official lists were available. It might have been the source of the present lists. I think they correctly show the decline of prices during the period, but during this period there was a five cents a can ferry charge, and I believe in most cases a freight charge, both coming out of the return to the farmer. I recall returns during the early summer months of the 1880's or 1890's showing fifty cents for a forty-quart can of milk, or $1\frac{1}{4}$ cents a quart. The minimum return that I recall was thirty-nine cents on one occasion for a forty-quart can of milk.

In 1897, when J. D. Gilmore and other promoters were trying

to organize a single corporation for the distribution of all the milk in the metropolitan market, they estimated that Orange County farmers alone lost $75,000 a year through defaults of individual dealers. I should consider that a conservative estimate. During the following fifteen years the defaults and bankruptcies of incorporated dealers developed into what we would now call a "racket." The losses have been reduced by licensing and bonding dealers, but losses have not by any means been fully eliminated.

CHAPTER III

LAWS, REGULATIONS AND INVENTIONS

Cows in Brewers' Stables.

The first hygienic milk problem of the New York City Board of Health presented itself in the brewers' stables of the swill-fed cows at the very beginning of the nineteenth century. But it was not until 1856 that the Brooklyn Common Council adopted an ordinance merely limiting the herd that could be kept on a city lot to three cows. Later that same year an ordinance provided that no more than three cows could be kept in any stable or inclosure or upon any lot of 25,000 square feet, between May 1 and November 1; no more than four on half an acre; or more than six on an acre; and not more than twelve on any lot whatever. A fine of $10 per cow was the penalty for violation.

A distiller by the name of S. T. Husted had influence enough to have a special meeting of the common council called to reconsider the terms of the ordinance. He succeeded in having it changed so that distilleries then in operation and milkmen then in the milk business were exempted. That was probably the first "joker" in an American milk law.

Two years later, 1858, a committee of the Board of Health heard witnesses testify that the stables were clean, the cows healthy and the milk unadulterated and pure, or words to that effect. The ordinance was repealed in an action which the city papers called "a whitewash." The action, however, stirred up the press and the public to such an extent that in 1861 a bill to prohibit the adulteration of milk and to stop the production of swill milk passed the State Senate but failed in the Assembly. The following year it passed both houses and became a law.

In the year 1866 the Department of Health was organized in New York City, and the swill milk business was then banned in the city. Brooklyn, however, was yet a separate city. As late as 1904 it was found that six thousand cows in that city were yet fed on brewers' waste. The law was later enforced there, so, after a century of wallowing in filth and corruption, the swill-fed cows disappeared.

Adulteration Practices.

The quality of the milk produced in filthy city stables from cows fed on brewers' swill was so poor that adding fresh pure water was said to improve it. The bluish color was modified by adding chalk and plaster of Paris and molasses. A charge of adulteration of milk against a dealer under the law of 1861 was dismissed because the judge doubted that the Legislature intended that mixing water with milk was included in the term "adulteration." An amendment to the law in 1864 removed that doubt. The amendment permitted the addition of enough ice to keep the milk cool, but any milk obtained from animals fed on brewery waste or any substance in a state of fermentation or putrefaction was declared to be impure and unwholesome.

In 1869 the Board of Health was not much alarmed over the fact that New Yorkers drank 40,000,000 quarts of water with its 120,000,000 quarts of milk. It estimated, however, with indignation, that the water cost the people who drank the milk $4,000,-000 annually. Six years later, in 1875, it was estimated by a Professor Chandler that adulteration netted the milk dealers, after paying all fines, $8,000,000 in two years.

In 1876 three city inspectors were appointed and armed with a lactometer to inspect milk and detect adulteration of milk by addition of water. This was the beginning of milk inspection. By 1880 two additional inspectors were added and an attempt made to examine all stores where milk was sold. Only one inspection a month was the best this force could do. At that time it was said that milk sold by one out of every five dealers contained from fifteen to forty per cent of water.

When the milk in the stores and on the streets had been fairly well inspected, attention was directed to the ferries at night. There a process of wholesale adulteration went on openly. Milk

LAWS, REGULATIONS AND INVENTIONS

was skimmed as it came on the boats from the trains. The skim milk was poured in with the whole milk and then four quarts of water added to fill the forty quart can. The ferrymen were in collusion with the milk men and gave the latter warning of the approach of the inspectors. An excuse was offered that the milk kept better when watered, and the processing helped pay for the wear and tear of the wagons and cans.

So long as the courts imposed only small money fines for convictions, the milk dealers joyfully paid the fines and took their profits. In 1895, however, the judges began to impose prison terms as well as cash fines, and that practically stopped those gross adulterations.

On August 13, 1893 the *Tribune* published a report of a meeting of milk dealers called by a dealer by the name of George Dainty, who acted as chairman. The meeting was held in Pythagoras Hall in Canal Street to protect both the law and the inspection. One of the excuses offered for pleading guilty to charges of adulterating milk was that dealers "sometimes" got adulterated milk from farmers and in their hurry they neglected to examine it. A resolution was adopted recommending that inspectors be appointed who "would treat milk men right."

The State Dairy Commission.

The State Dairy Commission was created in 1884. Josiah K. Brown was the first Commissioner. In 1893 the name was changed to the Department of Agriculture and Mr. Brown was succeeded by Frederick Schraub of Lowville, New York. Successive Commissioners since have been Charles A. Wieting, 1896; Raymond A. Pearson, 1908; Calvin J. Huson, 1912; Charles Wilson, 1916; Berne A. Pyrke, 1921; Charles H. Baldwin, 1932; Peter G. Ten Eyck, 1934; and Holton V. Noyes, 1937. George L. Flanders was appointed Assistant Dairy Commissioner June 1, 1894. He was appointed Assistant Commissioner of Agriculture by Commissioner Schraub and reappointed by Commissioner Wieting. Commissioner Pearson designated Mr. Flanders as counsel, who remained in that position until October, 1925, when he retired.

In 1885 Mr. Brown made a lengthy report of his first year's work. He estimated that 80 per cent of the traffic in unfit dairy products had been broken up. He had 60 indictments awaiting trial. However, the law did not prevent traffic in oleomargerine. Large quantities of it were being shipped into the state and through the state and to interior markets. He recommended that purity of milk standards be kept so low no one could complain. Normal milk, he held, should not contain more than 87.5 per cent water, 12.5 per cent solids, 3.2 per cent fat, and 9.3 per cent solids not fat. He defined adulteration of milk as caused by taking out fat and substituting water. Legal milk was later defined as containing not less than three per cent fat and twelve per cent solids.

Milk is adulterated in the eyes of the law when anything is taken out of it or anything put into it. Originally this prohibited the practice of syphoning milk from the bottom of the can or other receptacle after the cream has risen to the top, but this is not generally objected to now if the fatless milk is removed by mechanical means. Fore milking is also resorted to by some producers to meet the requirements of higher fat content. Standardizing by mixing milk of low fat content with milk of higher fat test is a general practice.

The Washington Square Pump.

I have been told by eye-witnesses of that early period that a pump in Washington Square, in New York City, was regularly visited by milk distributors' wagons in the early morning to fill up the empty space in the top of the milk can before starting on the delivery route. The churn or the chemist's expensive analysis was the only way to determine the amount of fat in milk. The addition of water was indicated by the density of the milk as determined by the lactometer. But a solution of salt "fooled" the instrument, and burnt sugar and other coloring materials corrected the blue color of watered milk.

Health Permits.

In 1896 the Board of Health adopted the following ordinance:

LAWS, REGULATIONS AND INVENTIONS

"No milk shall be received, held, kept, offered for sale, or delivered in the city of New York without a permit in writing from the Board of Health and subjected to conditions thereof." Later the same year it was made unlawful to mix any other substance with cream or to sell condensed milk containing less than 25 per cent fat content or with anything added except sugar.

The first mention of bacteria as applied to milk is found in 1896 in a report of the city Board of Health.

Nathan Straus opened a station for the free distribution of pasteurized milk in New York City in 1893. This was the beginning of pasteurization.

In 1896 there were 2,800 cows in New York City, 2,250 were producing commercial milk, and 550 for private use. In that year the Health Department adopted a system of stable inspection and used the T.B. test for the first time.

Three Inventions.

Three inventions of this period did much to hasten the development of the dairy industry. Gail Borden invented the vacuum condenser in 1856, which happens to be the year in which I was born. The De Laval separator was invented in 1878, and the Babcock tester in 1890.

During the 80's on my father's farm, milk from about 30 cows was strained after each milking into flat eight-quart shallow tin pans arranged on a rack in the cellar. When the cream had risen to the top, the womenfolk would skillfully remove the cream with a ladle. The skim milk and buttermilk were fed to the pigs and calves. The cream was stored in large crocks for sufficient time to develop the desired acidity. The temperature was then regulated and father's job every morning was to agitate the cream with the "dasher" until the "butter came." Later a swing-box Davis churn made churning easier. Later still a dog on an endless chain tread did the churning.

In the meantime the Stoddard Manufacturing Company produced a cabinet in which deep glass tubes were surrounded with iced water, with a clean arrangement for skimming the cream, and then drawing off the skim milk by gravity. The iced water separated the cream between milkings, kept the milk and cream sweet, and by a more perfect separation produced more cream, and a better quality of butter. The separator made a still further improvement in saving of time and labor. It made a still more perfect separation and more cream.

The Babcock test consists of dissolving the curd of milk in a glass tube with an acid without affecting the fat. By agitating the resulting liquid with a simple mechanical instrument and adding hot water, the fat being lighter than the liquid rises to the top. The glass tube is gauged so that the percentage of fat in the milk is easily read on the scale. It is so simple in operation that any boy can readily operate it from simple instructions.

Dr. S. M. Babcock, the inventor, was a native of New York State. He was born at Bridgewater, N. Y., on October 22, 1843. He taught chemistry at Barnard. He was a practical chemist at the Geneva Experiment Station from 1882 to 1888. He was a chemist in the Wisconsin Agricultural College in 1890 when he perfected the test. If patented it would have made him a large fortune, but he felt that he had developed it while in a public service and he wanted to give it to the public. It was a great contribution to the dairy industry.

The vacuum condenser made fortunes for manufacturers. It created a demand for milk in the communities where plants had been erected. The product became a factor in the industry, but the prices paid to the milk producers have not been commensurate with the profits of the manufacturers.

Greed for Milk Dollars, Then and Now.

My city milk recollections do not precede the year 1870. The records show, however, that New York City harbored cows and produced milk for practically the entire period of the nineteenth century at a staggering cost in money, morals and life. The stables were sunless, airless and dark. The cows never knew fresh air, rain or sunshine. They wallowed in dampness and filth. They were fed on swill. They grew sick and died in their narrow stalls. Their milk was acid and unfit for human consumption.

The wealthy brewers developed this system of production and the milkmen were their allies. Together they corrupted politicians and silenced officials sworn to enforce the laws which they chose to forget. Together they defied the protests of the people, the doctors, public spirit, and the best elements of the city press. Yielding slowly and grudgingly to a growing public demand, the brewers were forced to abandon the mass production of unwholesome milk, but their partners, the milk distributors, turned to adulteration of farm produced milk and other forms of trickery to increase their dishonest gains.

During the long period of this regime of filth and corruption in the city's milk production, the proponents of health and decency pleaded for the pure fresh milk of the farms. Country surveys were made by physicians and civilians and reports made of the availability of farm produced milk, and recommendations were made for its use.

I do not wish to minimize the work and influence of the sincere city health authorities, city officials, and city distributors who then and since helped in creating a cleaner and healthier system for the distribution of milk in the city. I wish to give full credit for the improvement to those, living and dead, who had a part in it. But to those officials and distributors, including some of our own official leaders, who, with an accusing glance of suspicion at the farmers, point to our high quality of milk and the saving of babies' lives as their sole accomplishment, I deem it proper to remind them of the record.

I call their attention to the fact that farmers were producing pure, clean, and healthful milk when the city itself was for nearly a hundred years producing and feeding its babies and people the lowest grade of milk on record anywhere. I would remind them too that the present-day racket in milk distribution is but a refined form of the system that prevailed during the nineteenth century, adjusted to the changed conditions; and that babies and children of the present time are suffering as a result of the same instincts of avarice and greed that dominated the industry during the last century.

Give Farmers a Chance.

I carry the comparison still further. I confess to a deep-rooted interest in the farm and to an abiding sympathy with the dairy industry, and especially the producers of milk. But I have had opportunity for intimate, almost daily, contact with the business for seventy years, and from my experience, I have come to the unprejudiced, deliberate and sincere conviction that, if left to their own initiative and pride and protected in their rights to negotiate a fair price for their milk, the dairymen would of their own volition have produced and delivered a better quality of milk than we have so far been supplied.

Left entirely to the self-interest and human instincts of farm men and women, I believe that as a whole the quality and supply of milk today would be just as superior to what we are now using, as the farm milk of one hundred years ago was to the city production of that time. With the natural and economic rights of the dairy farmers of New York State restored and protected, I believe they would be producing milk to the full capacity of the farm. That would mean from two to four times our present production. That in turn would mean greatly reduced cost of production, and such a low price to the consumer that milk and its products would be the principal item of food of the city population.

The same brand of ignorance, cupidity and greed that poisoned infants and children in the years of slop-fed cows, and starved infants in the days of adulterated watered-milk, is at this very time depriving infants and children of a full supply of high quality milk at a price within their reach.

I fully realize that there will be interested critics to jeer at this statement. Some will "laugh it to scorn," others will sagely treat it with silent contempt in the fear that public discussion of it would convince the public of its truth. These will repeat the whispering campaigns to discredit the author, or failing in that, to distract attention from their own vulnerable milk structure. Their predecessors used such strategy a hundred years ago when they sought to prolong illicit sale of poisonous and adulterated milk for human consumption.

Our present racketeers are more safely entrenched in an economic armour of monopoly. But they prey on a different generation. Their days are numbered. Some day in the future when dairy farmers market their own milk on their own terms, and city people eat and drink milk for the sake of both health and economy, some scribe will hunt up the record of today. He will tell a new generation how a middleman profiteering system enriched itself in the mid-twentieth century by establishing itself between producers and consumers of milk, whence they exploited the former and robbed the latter with a revised scheme then already a century old.

CHAPTER IV

THE GROWTH OF INDUSTRY

The Upward Curve.

The period of the first two decades of the twentieth century was an epoch in American life. From the demonetization of silver in 1873 to 1896, the curve of prices was downward. During that period population had increased from 35,000,000 to 75,000,000. Our real money was limited to the gold supply and the production of gold was low during the whole period. Money was scarce and dear. That meant low prices. The increase of population and industrial growth multiplied the demand for money far in excess of the supply. During the Civil War, 1860-1865, gold payments were suspended except for payment of duties on imports and interest on the public debts. Paper currency, greenbacks, were issued and the public debt increased. This inflation caused prices to rise, but deflation followed in the late 70's after the provisions to return to payments in gold were announced, and prices fell to a new low level.

In 1897 the production of world gold suddenly increased. It doubled and trebled in future years in an upward curve to 1923. This increase in the gold supply under our banking laws made it possible and profitable for the banks to issue national bank notes and extend credit to industry, thus increasing the volume of money in circulation. Under this stimulus the curve of commodity prices rose regularly up to the beginning of the World War, and under war conditions shot up to the peak prices of 1920 to an index of 225, or $2\frac{1}{4}$ times the 1910-1914 five year average. Under the deflation of that year (1920) the price index dropped to 150, or $1\frac{1}{2}$ times the 1910-1914 basis.

Business Trusts Emerge.

Up to about the middle of the nineteenth century agricul-

THE GROWTH OF INDUSTRY

ture was the chief industry of the American people. It was independent, individualistic, and self-reliant without interference or aid from government. Following the Civil War the nation turned to manufacturing, mining, oil refining, railroad building, insurance and banking. These corporations gathered up the savings of the people and the surpluses of fiduciary corporations, such as insurance companies of all kinds, and surpluses and credits of the national banks. As infant industries the manufacturers were protected from foreign manufacturers by protective tariffs. Soon these corporations merged or combined with one another to form great trusts to fix and maintain prices.

By the beginning of the twentieth century we had the coal trusts, the oil trusts, and numerous manufacturing combinations. Among the railroads we had the Hutchinson system of the Southwest, the Morgan-Hill system and the Harriman system, both of the Northwest, and the Vanderbilt system in the East. We also had the Morgan and Rockefeller groups of banks. All of these and more were intermingled and interlocked in the corporate trust systems estimated as ranging in capital from a few million to twenty billions of dollars. In all instances, these interests were controlled by a comparatively small group of men who had lost their sense of proportion. They controlled not only the trusts and industries but the Federal Government. As they progressed, money lost its lure as a convenience to supply wants. Power had become their ambition. Money was the pawn on the checkerboard of their games.

The Morgan-Rockefeller group held 341 directorships in 112 corporations. The United States Steel trust was organized in 1901 with a capital of $1,100,000,000. It combined 5300 formerly independent plants. Not long after, Andrew Carnegie's share of the annual income from it was $25,000,000. At his death in 1877, Cornelius Vanderbilt left the major part of his $100,000,000 to his son William Henry, who succeeded his father as president of the New York Central and Hudson River Railroad. When the latter died eight years later as head of the "Nickel Plate" system in Chicago, he left $200,000,000. It is said he boasted of being the richest man in the world. "The public be damned" is a frequent quotation imputed to him.

When a memorable coal strike was under way in 1902, in an effort of laborers to correct intolerable conditions in the coal mines, George F. Baker declared that "God in his wisdom had given the control of the property interests of this country to Christian men who would protect the interests of laboring men." Mr. Baker expressed the attitude of this group of industrial and financial leaders at the time. They had gained so much power that they saw themselves as industrial kings by divine authority and right. As far as they were concerned, American principles and ideals as expressed in the Declaration of Independence and basic laws of the country had gone out of the window.

Agriculture vs. Industry.

In 1790 nine-tenths of the people lived on farms. Farmers were the leaders and substantial people of the nation. They cleared the forests, built the roads, the schools, the churches and the institutions. By 1890 only three-tenths of the population were farmers. Forty years before, in 1850, they owned half the wealth. In 1890 they owned only a quarter of the wealth. At the latter date the frontier closed. The urge to the free open plains of the west had broken through. The fertile virgin soil of the prairies had flooded the markets of the East and of Europe with cheap products of the land. Corn was used for fuel. I recall a letter in the early 90's from a Prairie State farmer offering to ship a carload of horses to the East if anyone would pay the transportation charges.

Farmers began to feel that something was wrong. They had been told that high tariffs were necessary to protect infant home industries and the wages of labor. It was argued that industry would create a home market for farm products, and that the competition of one factory with another would keep down the cost of manufactured products. But the price of the products raised by farmers would hardly pay the freight. The wages paid labor would hardly support the families. The attempt of labor to form unions was opposed by employers, and strikes were defeated by

THE GROWTH OF INDUSTRY 31

employment of foreign labor and armed soldiers. Instead of the promised competition there were mergers and trusts to control prices.

Agriculture had become second to industry and trade in numbers and in accumulated wealth. But it had fed the nation and produced the raw material for industry. It was outwitted in exchange but it maintained its freehold and kept alive its cherished American ideals.

Farmers of the Central West resorted to politics. They organized the Farmers' Alliance. In 1890 they elected three United States Senators and fifteen Representatives to Congress. As the People's Party they adopted a platform demanding concessions which they deemed essential to their welfare. There was nothing in it that would now be considered radical. In fact, most of the provisions which were ridiculed and denounced at the time have since become law. But the howl of the groups in power was loud and alarming. The farm revolt subsided for a time to break out six years later under the leadership of William Jennings Bryan.

A False Price Parity.

While the prices of industrial goods increased far beyond the modest price increase of farm products, farmers shared to some extent in the general prosperity of this period. The middlemen, however, multiplied and exacted increased shares. This left labor and service unprotected. About the beginning of the second decade of the twentieth century this condition gave rise to the complaint of the "high cost of living." I recall a large agricultural meeting in the Capitol in Albany which had been called to do something about it. Farm leaders, educators, industrialists, and transportation interests were there. The head of a manufacturer's organization of Rochester, N. Y., stated the case simply and clearly. Up to that time operators, he said, had been satisfied with their wage. Of late their cost of living had increased. They had no surplus to fall back upon so unless the cost of food could be reduced, the factories faced an increased pay roll.

There was no suggestion to pay the increased labor cost out of the increased price of the factory products. The remedy centered on the problem of reducing the cost of food at the farm. Educational work was recommended. Model farms run scientifically were suggested, so that farmers could have an object lesson of how to do it and produce at less cost.

President Brown of the New York Central Railway volunteered to set an example. He promised then and there to operate three such model farms in three different parts of the State under the auspices of his railroad. True to his promise, he did so. The papers got news releases as the farms were bought and equipped with material and workers. One expert was engaged to supervise the three farms. The optimism ran high. During the growing season the releases recorded some discouragements. The first harvests were a bit disappointing. The releases grew less frequent. At full harvest they became discouraging and finally stopped coming. If farmers really learned anything from the experiments, it was probably how not to do it. No release reached us admitting that a farm could not pay current wages for all work and operate at a profit. But that is what was demonstrated by Mr. Brown's experiment.

At the time I felt a sense of discouragement from the fact that for forty years since 1870, farmers had been struggling with desperate conditions and yet were at a price disadvantage in comparison with industrial products, and that at the first intimation of partial relief an agricultural meeting could find no solution of the high cost of living except a reduction of the meagre income of the farms of the State.

It was in further discussion of this problem of the "high cost of living" that an industrial meeting in the city of Binghamton provided for the first County Farm Agent, which soon developed into the Farm Bureau. For a time the industrial and commercial interests contributed to the cost, for which some of them demanded consideration. Later the expense of the Bureau was assumed by national, state, and county appropriations, and to some extent by individual farm contributions. Other industries employ research workers and experts at their own expense. They treat this expense as part of the cost of production, and add it to the price of their product.

CHAPTER V

THE O'MALLEY INVESTIGATION

The 1910-1914 Index.

Agricultural conditions had been so bad after the 1929 slump that the period just before the World War apparently seemed good in comparison. Economists have taken average prices of products for the five-year period, 1910-1914, as a basis for index numbers to indicate rise and fall of prices since. One of the political parties jumped at the conclusion that if prices in that period were again reached it would solve the farm problem. In its party platform of 1928 it promised to maintain that parity. As a matter of fact it was no just parity.

During the 1910-1914 period farmers' prices were not on a parity with industrial prices. In industry the wages of skilled labor and the salaries of managers were higher than wages of skilled farm labor. The income on industrial capital was more than the income on farm capital. The profits of industry were higher than the profits of the farm. To perpetuate that state of affairs by law, as was the boast of the Agricultural Adjustment Administration (AAA), would be a rank injustice to agriculture and a violation of the Constitution, as was demonstrated by the unanimous decision of the United States Supreme Court.

There was not a man who touched a farm product from the time it left the farmer's wagon until it reached the consumer's door, who did not get more out of it in proportion to time and energy devoted to it than the farmer got who produced it. This included the car-loader, engineer, fireman, brakeman, conductor, city truckman, commission merchant, wholesaler, retailer and even the delivery boy. The things farmers had to sell were not on a parity with the things farmers had to buy. During the period farmers were in protest against the discrimination. It was against such prices that dairymen fought in 1916. To adopt that parity of prices permanently would be to continue the discrimination and leave dairy farmers no hope of ever being permitted to determine a fair price for themselves.

While general prices rose steadily under the influence of the increased gold production from 1896 up to 1914, prices of agricultural products lagged far behind industrial goods. After thirty years of falling prices an upward turn came with a grateful feeling, but the increase to the farm was never commensurate with the increased price farmers had to pay for household supplies and production needs. It was during this period that *The Rural New-Yorker* found that farmers' returns did not exceed thirty-five cents of the consumers' dollar. The estimate taken from a comparison of farmers' "account sales" with prices of farm products in the markets, was at first disputed and reviled as unscientific. It was later amply confirmed.

Prices and Profits in 1909.

In the meantime, Borden's Condensed Milk Company had secured a substantial footing as a fluid milk distributor in the City of New York. It made a practice of fixing and announcing a schedule of monthly prices to be paid producers for milk, six months in advance, beginning April 1st and October 1st each year. The Borden Company made the boast of providing a market for all the milk furnished at its prices. So well it could. The price was estimated on the value of the milk for manufacturing purposes, with a good profit on the manufacture and a liberal margin for safety. At that time winter production was light and spring flow heavy. To encourage an increased winter production, the price for these months was slightly advanced, and to even up costs the schedule of prices for the spring months was proportionately reduced below the fair value of milk for by-products.

For a considerable time the Consolidated Milk Exchange continued to announce prices, which varied little from the Borden schedule, and finally the low Borden prices became so attractive and satisfactory to all city dealers that their schedules were accepted by practically all distributors. As general business condi-

THE O'MALLEY INVESTIGATION

tions improved successively for the first sixteen years of the twentieth century, a number of substantial concerns developed a good city trade. Frequently these competitors would pay a bonus of 10 to 30 cents above the Borden price; but in the flush seasons the producer could only accept the Borden price or sell the cows. This method of announcing milk prices ended in 1916.

On December 6, 1909, having received complaints that combinations in restraint of trade and a monopoly in the milk traffic existed in the City of New York, Hon. Edward R. O'Malley, Attorney General of the State of New York, conducted an investigation before a referee appointed for the purpose by the Supreme Court. His report was submitted to the Legislature on April 20, 1910. Brief reference has already been made to this investigation. The specific complaint was that the dealers, as members of a bogus exchange, had agreed among themselves to fix the maximum price to be paid farmers for their milk. The hearing began December 6, 1909 and continued until March 3, 1910.

Testimony of many witnesses, including dairy farmers and milk dealers, was taken and condensed for publication in the report. Evidence was adduced to show that in the latter part of 1907 the dealers increased the price to consumers from seven to eight cents a quart. In 1908 the price to consumers was increased to nine cents a quart. Owing to widespread criticism, the price was reduced back to eight cents early in 1909 but on November 1, 1909 dealers again increased the price to consumers to nine cents a quart, with no increase to producers on either occasion. Evidence taken on February 28, 1910, during the investigation, was furnished to a grand jury which found individual and blanket indictments against members of the board of directors of the Consolidated Milk Exchange, Ltd.

The testimony revealed that Borden's profits in New York and Chicago for the year 1909 over 1908 increased $340,353.12, and that its total profits for the year were $2,617,029.40. Its capital invested in New York was $4,890,487.00 and its New York profits alone were $682,367.16, or 14 per cent net profit, which was about $170,000 more than the profit for the previous year. The capital stock of the company was $25,000,000, of which $15,-

428,408.46 was issued for trade marks, patents and goodwill—wind and water.

Sheffield Farms, Slawson-Decker Company, organized eight years before, earned net for the year ending February 28, 1909, $221,694.63, and for the eight months ending October 31, 1909, earned net after paying all expenses, $257,923.47, which was a net profit of 120 per cent on the original $200,000 investment.

On October 31, 1909, a remarkable statement appeared in the *New York Times* and other city papers. It said: "Milk up a cent; now nine cents a quart. Farmers raised price. New contracts signed November 1 give the producers a half cent a quart over last Summer, making the price 4½ cents a quart." The exact prices quoted by Borden's and the Consolidated Exchange in the 32-cent zone for the previous summer were, per hundred pounds:

	Borden	Exchange
June	$1.00	$ .99
July	1.15	1.00
August	1.25	1.24
Average	$1.13	$1.08

From this was deducted 32 cents freight to Jersey City, and 5 cents ferry charge, leaving the net 76 cents and 71 cents respectively. So the net price to these producers would be about 1.5 cents a quart instead of 4½ cents as reported to consumers. In any case, the net average gross price to producers for the three summer months was 2.5 cents. Not only so, but the Borden average for the whole year was $1.52 per hundred pounds, which after taking out freight and ferry charges left only $1.25, or 2.6 cents a quart. In the government lists of that year, 1909, the quotation is 3.38 cents a quart, or $1.59 per cwt.; and part of the year 1909 is in the so-called parity period.

The most significant thing about the incident, however, was a copy of an agreement published in the *New York World* which that paper said was made between the *New York Tribune* and the milk dealers to conduct an "educational campaign." The alleged agreement said in part:

"We, the undersigned, engaged in the milk trade, and naturally desirous of promoting the best interests of the industry, have consented to co-operate with the *New York Tribune* in the pursuance of an educational campaign to be conducted in the columns of that influential journal, to the end that both public and press shall be set right on the vital question at issue between it and them: and for the purpose of sharing the expense of the articles do hereby subscribe the sums set opposite our names."

According to the figures the *Tribune* was to receive $5,000, or $1.00 per line for the "service." The plan was to convince city consumers that the price they paid for milk had been too low, and that in view of the farmers' demands for a higher price the dealers had no choice but to increase the retail price. The publicity by the *World* may have thwarted the raising of the "campaign" fund.

State Probes Milk Monopoly.

The O'Malley Report recited the organization of the Milk Exchange, Ltd. in 1882 and the annulment of its charter by the Supreme Court in 1895, as the result of an action against it by the Attorney General.

The report also cited the organization of the Consolidated Milk Exchange, Ltd. a few weeks later under the laws of the State of New Jersey substantially by the same men, and for practically the same alleged purpose. It filed a certificate to do business in New York and other states. This certificate was voluntarily withdrawn during the time of the O'Malley investigation, but the members continued to announce prices up to June 1913, when the laws of New Jersey intervened.

The Report explained the method of the Consolidated Milk Exchange in fixing prices as follows:

"Each director was a member of the 'Committee on Values.' After an informal discussion of what the price of milk ought to be for the succeeding month, a ballot was taken, and as a result of that ballot the price was fixed. A sample of the action of the 'Committee on Values,' as disclosed by the minutes of the·board of directors, is as follows:

'A recess was then ordered, subject to the call of the chair, and, on reassembling, the chairman on Values reported that, in the judgment of the committee, they found the value of milk to be _____ per can of forty quarts, less freight charges from each respective shipping-point, together with an allowance of five cents per can for shortage. This report was duly accepted.'

"The word 'value' is used instead of the word 'price' for the undoubted purpose of evading a violation of the law prohibiting the fixing of prices. Indeed, a record of one of the meetings of this company shows that the word 'price' was originally used, but was stricken out and the word 'value' inserted.

"Borden's Condensed Milk Company, the largest dealer in milk in the State, announced every six months in advance the price it would pay producers for milk during the ensuing six months. Borden's prices and those established by the Exchange were not always the same, but on the average were substantially the same. In the same way the Sheffield Farms, Slawson-Decker Company, the second largest dealer, sent out the prices it would pay to the producers. In some cases the price was exactly the same as Borden's; in other instances, the same as the Exchange, and in others a price which approximated Borden's and the Milk Exchange prices.

"The result was that a producer desiring to sell his milk in the New York market was compelled to sell either at Borden's or the Exchange prices, which were practically identical, and if not satisfied with either of these, he was compelled either to manufacture his milk into butter or cheese, or market it with unknown and oftentimes irresponsible dealers."

As a convenience for fixing prices to the consumer, the Milk Dealers' Protective Association was organized under the membership corporations law in New York at the time the Consolidated Milk Exchange was organized in New Jersey. The evidence tended to show that the members all endeavored to maintain a price fixed by themselves to stores and consumers. The Report continued:

"If an independent dealer, not a member of the Association, attempted to sell milk at a lower price than that established by the Association, what was known as the 'dead' wagon was started after him. The peculiar duty of this 'dead' wagon was to go around to the customers of the independent dealer and to offer them milk at a lower price than the independent was selling at. This 'dead' wagon was maintained and supported by the Milk Dealers' Protective Association. If the operations of the 'dead' wagon were not successful in putting an independent dealer out of business, an attempt was usually made to cut off his supply of milk by coercion, threats or influence exerted upon the party who was supplying the independent with milk, sometimes as high as $1,500 being offered to the party supplying the independent with milk if he would break his contract with the independent or send the independent sour milk for a few days.

"General competition among dealers as to the price at which they will furnish the consumer with milk has ceased. They all put the price up at the same time to the same amount and all, with the exception of one company, put down the price to the same amount at the same time. The consumer is at the mercy of the dealer; he must buy milk at the price established by these dealers or go without it."

Attorney General O'Malley stated in his Report that "in his judgment there existed in New York City a condition which in effect is a combination which fixes the price at which the producer is obliged to sell milk and that he has no voice in determining what that price will be."

CHAPTER VI

MILK PRICES AND GRADES

The Borden Formula.

The Federal economic bureaus estimate the annual average price of milk back to 1868 or longer, but the figures can be taken only as an estimate. That, however, is true of all our milk price statistics. The accountants have no doubt done the best they could with the figures and facts available. These are taken largely from dealers' quotations and schedules, but in practice the figures are changed and modified. There were no fixed schedules up to 1882.

The Milk Exchange, Ltd. then sent out prices arbitrarily fixed until 1895, except for the first year when farmers negotiated the price as a result of their victory in the Orange County Milk War. From then on to 1913 the Consolidated Milk Exchange, Ltd. issued prices. Borden's, originally a condensed milk manufacturer, contracted more or less regularly for its milk supply. The company was not a factor in the fluid milk trade until 1900 or later. About that time, however, it began to issue a fixed schedule of monthly prices for six months in advance, from April 1 and again from October 1.

Borden's schedules were seldom modified during these six months' periods. After they got well established in the fluid trade, the smaller dealers followed Borden's schedules. This was especially the practice after the Consolidated Exchange, Ltd. closed up in 1913. But, as a rule, the better class of the smaller dealers paid a premium on the Borden prices. During the periods of short supply, Borden's was obliged at rare times to increase its schedule five to ten cents a hundred pounds.

The company had a formula on which it estimated its prices monthly in advance for the six months' periods. Its prices were based on its estimate of the prices of butter, cheese and condensed milk. While milk was flush in the three Spring months, Borden's dropped 15 cents a cwt. under the manufacturing price.

Four things helped the company gain a dominant position in the market: (1) Its price was known definitely in advance; (2) it paid promptly; (3) it induced farmers to help finance a comparatively large number of its country plants, supplied shipping cans and led farmers to abandon butter and cheese-making, so that when prices dropped the farmers were helpless; and (4) the price was so low the other dealers were inclined to pay the company's price and no more, except in case of necessity.

The members of the Exchange, for the most part, attempted and succeeded in contracting at less than the Exchange price. Ten to fifteen cents a can less was the rule. Their practice was to go back into the remote districts and induce the butter and cheese patrons to go into the fluid milk market. Frequently they made deductions in the returns alleging that the milk was sour, or that the cans were not full to the top, or that the cans were dented. Some of them contracted the habit of holding up payments. Credit losses to the producers were heavy. Finally the State law required the dealers to take out a license and put up a bond to secure payment, at least in part. This reduced the failures, but the bonds seldom have covered the full credit and the losses have been too frequent and too heavy even down to the present time.

Paying on Fat Content.

After the Babcock tester provided an easy and inexpensive way to determine the butter fat content of milk, the dealers began to base the price on the amount of fat in the milk. First they made a low flat price and then offered a premium of three cents a point (one-tenth of one per cent) on milk testing more than 4.1 per cent fat. For a time in 1912, Sheffield Farms paid this premium for milk testing more than 4.5 per cent fat. This mattered little because the bulk of the milk did not test over 4.5 per cent fat. Later, the base was 3.8 per cent, and finally the price was based on 3 per cent milk, the standard fixed by law, and three cents above the base price for every extra point of fat in one hundred pounds of milk. A large volume of milk, however, continued to be sold as "just milk" as it came from the cow up to about the time of the World War.

For a considerable time during this period complaint was made that the dealers separated milk, sold the cream to consumers, and then mixed the skim milk with the whole milk to reduce the fat content of the mixture as close as possible to the legal standard of 3 per cent fat. They had a formula worked out so they could make this mixture by first testing the whole milk to determine how much skim milk could be safely added. The formula was shown in a diagram and witnesses in the O'Malley investigation of 1910 reported that they had seen the milk separated in the country plants and had watched the different pipes discharge the milk into the separate receptacles. One witness testified that he understood his dealer had a separator set so that it would take out fat and leave 3 per cent fat in the milk.

In 1916 the Wicks Committee examined the records of up-state plants. The weight of milk and cream shipped out of the plant equalled the amount received from farmers but there was no record of skim milk shipped. As I recall it, there were some direct admissions that the skim milk was mixed with whole milk shipped to the New York market.

The following year I bought samples of milk all over the City of New York through the Department of Foods and Markets and had it tested with the Babcock test. Some of it was bottled milk and some of it was loose milk. Some of the samples tested as low as 2.7 per cent fat, and none of the samples tested showed more than 3.2 per cent.

It was generally understood and often admitted that, despite the law, the fat in milk was regulated. In fact, my information is that the practice is continued to some extent down to the present time. The law defined as "adulterated" any milk from which any substance had been taken or to which anything had been added, or that contained less than 3 per cent of butter fat or more than 88 per cent water. This allowed the mixing of milk rich in fat with other milk having a low fat content, but it did not permit anything to be added to or taken from the natural milk as it came from the cow.

Grading the Milk.

For a time after the Board of Health of New York City began to regulate the hygienic quality of milk about the beginning of the present century, some dealers regulated the price by the bacteria count of milk. This however, did not reach important proportions until January 1, 1912, when the city created three classes of milk; Grade A, Grade B and Grade C. It required that Grades A and B must be pasteurized. Grade C was allowed to be sold only for cooking purposes to hotels, restaurants, bakeries and manufacturers of milk products. Later Grade C was abolished, but it was not until 1926 that pasteurization of all milk sold in the City of New York was strictly enforced.

The extra price to consumers for Grade A milk was first two cents above Grade B. The premium paid by dealers to farmers for the extra cost of producing Grade A milk was 15 cents, 25 cents and 35 cents a cwt., depending on the number of bacteria colonies in the milk as shown by the test. At times the higher premiums were suspended, and for some months no premiums were paid, but the consumer paid her full two cents extra without interruption. The dealers made their own count. The high limit under the city regulations was 200,000 bacteria count. The higher the farmer's bacteria count, the less his premium, if any, but the dealer's profit was higher on that farmer's milk.

During the World War the price of Grade A milk to the consumer was increased to three cents a quart above the Grade B milk. This price to the consumer was definite and fixed. The premium over Grade B to the farmer was elusive and uncertain. The city required Grade A milk to test only 3 per cent fat, the same as Grade B. Originally the farmer's premium for Grade A milk was separate and distinct from the fat test. Later the bacteria premium and the fat differentials were combined in a single schedule as follows:

	(———Bacteria Colonies per C. C.———)		
Butterfat	(10,000 or less	10,000 to 25,000	Over 25,000)
3.0.....	$0.10	$0.06	...
3.1.....	.15	.075	...
3.2.....	.20	.10	...
3.3.....	.25	.15	...
3.4.....	.35	.20	...
3.5.....	.40	.25	...
3.6.....	.45	.30	...
3.7.....	.50	.35	.04
3.8.....	.55	.40	.06
3.9.....	.60	.45	.08
4.0 and up	.65	.50	.10

Still later, the dealers made the reward more illusive by refusing to pay a premium on Grade A milk testing less than 3.7 per cent fat. The premium was graded up, according to fat test, from 3.7 per cent to 4 per cent but no higher for milk testing more than 4 per cent.

On October 1, 1939, two of the largest dealers in the city market raised the base from 3.7 per cent to 3.9 per cent. But no matter what the fat test may be, the premium is highest when the bacteria count is 10,000 or less. Next highest is when the count is between 10,000 and 25,000, and the lowest when the count is over 25,000. The New York Board of Health required a minimum 100,000 bacteria count and 3 per cent fat. The dealer limited the farmer to a maximum of 25,000 bacteria count and 3.7 per cent fat.

Dealers' Profits in Grade A.

The producer's milk might test 4 per cent fat, but if the count exceeded 25,000 and less than 100,000, his Grade A premium was only ten cents per cwt. The basic price of Grade A milk was the same as for Grade B. The premium was an allowance for the farmer's extra work and expense in complying with the drastic hygienic regulations required by City and State for the production of Grade A milk. While the dealer has some extra precautions in the plant, the farmer assumed most of the work and expense in the Grade A milk supply. What his average premium may be is not known, but with a possible average of ten cents a hundred pounds to the farmer on all milk, the extra three cents a quart for the run of the tank looks big to the consumer. Of course, all of this milk with different tests and counts went into a common vat, and the consumer paid her three cents a quart for all of it.

These were not all the tricks in the Grade A prices. If the dealer reported that he had a surplus of twenty-five per cent of Grade A milk, the rules permitted a reduction in the returns to the producer. The farmer had no way of checking the accuracy of the surplus allowance. He was obliged to take the dealer's word for it as well as for the weight, the fat test and the bacteria count.

Although the Grade A producer was required to comply with expensive regulations not required of the Grade B producer, I have known milk from Grade B herds to be diverted to Grade A plants with no extra requirements whatever. Only recently a responsible dairy farmer has told me that he followed a "co-operative dealer's" wagon loaded with milk from a Grade B plant direct to a Grade A plant and had seen the Grade B milk delivered in the Grade A plant.

Here then is good ground for the consumer to complain that she pays three cents extra to give her children, perhaps her sick child, or in many cases her whole family, the best pasteurized milk, but may get Grade B milk which she could have bought for three cents a quart less.

I do not charge that all dealers resort to these tricks. I would not suspect that all did, but as far as I know, all have agreed without protest to the new plan of combining fat test with bacteria count as a base for Grade A premiums. The price for fat is based on food value. The Grade A premium is pay for care, labor and expense of keeping milk clean, cold and pure. These are different things. Combining them serves no one but distributors, and affords dealers an opportunity to cheat both producers and consumers.

Protest by Farmers.

In 1936 producers of Grade A milk at Washingtonville, in Orange County, New York, were protesting Borden Company methods of fixing premiums and some other procedure at the local Grade A plant. They expressed their sentiments in the following resolution which was unanimously adopted in the month of March, 1936. It was published at the time in *The Rural New-Yorker.*

"The following protest and appeal was approved at the annual meeting of the patrons of Borden's Grade A plant at Washingtonville, Orange County, N. Y. It was signed by the patrons and sent to the Dairymen's League office:

"We, members of the Dairymen's League Association, Inc., and patrons of the Borden plant at Washingtonville, N. Y., feel that the present schedule for paying Grade A premiums to Grade A producers is an injustice to the producer and works to his disadvantage, while at the same time works greatly to the advantage of the distributor. We feel that it is a great injustice to the producer to have the butterfat premium and the bacteria premium tied together.

"We feel that the maximum count of 10,000 bacteria per c.c. is entirely too low, since the distributor is allowed 100,000 count prior to pasteurization. If the distributor allows contamination to get into the milk, or bacteria that is already there, to grow more than 90,000 after receiving the milk, it shows laxity on their part.

"Therefore, Be It Resolved, That we strongly urge our board of directors to take this matter up with our buying dealers with a view of working out a schedule for premium payments that has some justice in it for the producer as well as compensating him more fully for his extra expense and effort. To produce Grade B milk at Grade B prices is bad enough but to be forced to produce Grade A milk at not much better than Grade B prices is ruinous.

"Therefore, Be It Further Resolved, That we strongly urge our directors to insist that the butterfat premium and bacteria premium be separated and that each be paid for on its own merits:

"That the bacteria count be raised from a maximum of 10,000 to 25,000;

"That if there is a suggestion on the part of the buying dealer that the butter fat test be raised above 3.7% or that we subject our dairies to more than four veterinary inspections a year, this suggestion be answered by our directors with an emphatic 'no';

"That once more we impress upon our directors how disturbed we feel, and how much we dislike the idea of having money deducted from our milk check to pay the salary of the veterinarians in the employ of our buying dealers, and working for the interest of our buying dealers;

"That a copy of this resolution with signatures be published in the Dairymen's League News."

Producers in other sections expressed similar sentiments.

In the year 1938, an announcement was made of a combination of some New York and Philadelphia Grade A milk dealers including Borden's Farm Products of New York, Sheffield Farms and Dairymen's League Cooperative Association, Inc. of New York, Abbott's Dairies and the Interstate Milk Producers' Association of Philadelphia, for the alleged purpose of the improvement of the quality of the product. On the authority of past experience it is safe to predict that the extra quality, if any, put into it would be at the expense of the producer without increasing his return, and that the profits, if any, would be absorbed by distributors.

Health and Pasteurization.

In 1907 the Milk Committee, a voluntary group of citizens, reported that the City of New York received its milk supply from 30,000 to 40,000 cows in six States. The City Health Department after six years of country visitations had inspected 20,000 farms, and had begun the inspection of country plants, of which there were 670 in the territory. Few farmers at that time shipped direct to the city. The milk generally was delivered to plants, mixed, cooled and shipped in cars at a temperature of fifty degrees.

This was the time that scientists and medical men began to make practical application of the discoveries of Robert Koch in Germany and of Louis Pasteur of France. Koch discovered the specific germ (*Tubercle bacillus*) which causes tuberculosis. He believed, however, that bovine tuberculosis and human tuberculosis are distinctly different diseases. The Milk Committee advised the Mayor that the risk of transmitting tuberculosis through milk from cows to man was slight. The Committee believed that the danger was greatly over-estimated but advised systematic inspection and condemnation of cows revealing tuberculosis on physical examination.

Pasteur's experiments proved that fermentation and putrefaction are caused by germs present in the atmosphere. He then developed how these micro-organisms could be controlled by cold and heat and serums to cure and prevent certain diseases in man and beasts. Some of these organisms are harmless, even useful, others cause virulent disease. One of his processes of treatment involved heat. In the treatment of milk and other liquids for food and drink it has come to be called pasteurizing, or doing as Pasteur did. Pasteur did his work during the latter half of the nineteenth century. He died in 1895. The research and demonstrations of these two scientists, Pasteur and Koch, form the foundation of our system of milk hygiene.

As the system began to develop about the beginning of the present century, medical men and practical scientists realized that the new regulations would increase the cost of production at the farm. In 1909 Dr. E. J. Lederle, Commissioner of Health of the City of New York said, "The matter of an increase in the price the farmer must receive for his milk must soon be settled; it will become more and more urgent." He then quoted bottled milk selling retail for seven, eight and nine cents a quart, and "loose" milk retailing from stores for from four to seven cents a quart.

Again Dr. Lederle said, "A very necessary corollary to make these new conditions economically successful, and otherwise they can have no permanency, is that the public must be educated to the appreciation of their value. It will inevitably result in a higher price paid to the farmer and a general advance in the cost of milk to the public."

Dr. Lederle's prediction of an increase in the cost to the public was followed promptly and consistently up to 1940. The farmer has never been paid for the labor and expense necessary to comply with the sanitary regulations imposed on him from time to time by city and state regulations.

Less Milk and Lower Prices.

The adulteration of milk; the sanitary rules and regulations, both essential and fanciful; the cost of new sanitary equipments and supplies; the extra costs due to the changing system of distribution; the losses from defaults and bankruptcies; the advances in freight rates and other economic burdens, including taxation since 1840, have all tended to increase the cost of production and to reduce the New York farmer's net income for milk in his own State markets. Government figures quote an average of 4.6 cents a quart or $2.16 a hundred pounds as the average price for the year 1870.

The consumer's price in New York was as low as four to five cents for loose milk in stores and six to seven or eight cents up to 1907 for bottled milk delivered to the house. It went up to nine cents in 1908 and to 10 cents in 1916. These were basic prices for Grade B milk. Grade A was two cents higher. There was and is also an extra cent or two per quart when sold in pint bottles. In no year from 1900 up to the beginning of the World War did the farmer receive anything approaching the 1870 average yearly price of $2.16 per hundred pounds, or 4.6 cents a quart.

The New York farmer lost not only in a reduced price in later years but also in the loss of his markets because of the shipment of out-of-State milk and cream to the metropolitan and other markets. In 1917 the Wicks Committee Report contains the following summary and table (page 50).

According to the Wicks' Committee Report, in the year 1902 New York City consumed 12,665,922 forty-quart cans of milk and milk equivalent in cream and condensed milk. In 1916 the consumption was 24,459,663 cans. The increase in consumption between 1902 and 1916 was 11,783,741 cans or a 93.3 per cent gain, while during the same period New York farmers lost 18.7 per cent per dairy in production.

The same authority gives the population of the State in 1905 as 8,067,308 and in 1915 as 9,687,744, an increase of 1,620,436. It gives the cow population in 1900 as 1,501,608 and in 1916 as 1,301,754. This is a decrease of 199,854 cows, or 13.3 per cent.

In 1900 the general price level of all products was 82 per cent of the 1910-1914 prices. In 1916 the price level was 125 per cent, showing an average price increase in all commodities for the period of 43 per cent. For the same period the price of milk increased only 21 per cent to the farmer, but the dealer's increase to consumers was 38 per cent. Moreover, it was during this period that cost of production was increased for the farmer by the new hygienic regulations, so that in reality there was little, if any, net increase in the price of milk to the farmer for the period from 1900 to 1915. There was a distinct loss since 1870 when farmers set the price for the sale of their own milk. With that sole exception of the short period following the Orange County milk war in 1882, already noted, milk dealers had fixed the price of farmers' milk for forty-five years. The record is a reflection of dealers' domination of the dairy industry.

"STATEMENT SHOWING APPROXIMATE DECREASE IN POUNDS OF MILK YIELDED PER DAY PER DAIRY IN NEW YORK STATE FOR THE PAST 15 YEARS"
(From Wick's Committee Records)

Year	Average pounds per day per dairy
1902	278
1903	282
1904	283
1905	283
1906	266
1907	240
1908	235
1909	235
1910	255
1911	261
1912	246
1913	238
1914	233
1915	227
1916	204

Decrease in average pounds per day per dairy, 1902-1915—52. Per cent of decrease −18.7 per cent.

CHAPTER VII

MILK FREIGHT RATES

In 1842, when milk was first shipped by rail to New York City, the freight rate was one-half cent a quart. According to the records it was increased to 1 cent a quart in 1857 and further increased to 5½ cents a gallon in 1862. This was the time the Harlem Railroad began to carry milk to the city. By 1879, with the rate substantially the same, the Hudson River, the New Haven, the D. L. & W., the Erie Railroad and the New Jersey railroads were delivering milk to the city. There was also an extra charge for ferriage. Yielding to the threat of a legislative investigation, all the roads agreed to a reduction to 4½ cents a quart.

In 1884 the 40-quart can rate was reduced to 27½ cents in the case of milk and 45 cents for cream; but in 1885 it was again increased to 35 cents for milk, cream remaining at 45 cents. In 1890 the milk rate was reduced to 32 cents and cream to 42 cents a 40-quart can, but in 1892 the cream rate was increased to 50 cents a can. Up to May 15, 1897 these rates applied to milk whether in cans or bottles, and without reference to distance. The traffic had been extended to more than two hundred miles.

The Interstate Commerce Commission was established February 4, 1887. On April 23, 1887, the Orange County farmers filed the first complaint against the Erie Railroad with reference to milk rates from Orange County points to New York City. Producers in that section shipping 28 to 87 miles objected to paying the same freight rate as was imposed on milk shipped 183 miles. The Commission decided against them. Ten years later the same rate was extended to points 417 miles from New York City. The rate included bottles in cases as well as milk in cans. Complaints were again renewed, and the Interstate Commerce Commission reversed its decision and admitted the injustice of a single rate regardless of distance.

It then fixed the rates according to the following table:

STATEMENT OF ZONES ESTABLISHED AS A BASIS FOR FREIGHT RATES ON MILK DELIVERED TO NEW YORK CITY MAY 14, 1897

	Milk—40 Quarts			Cream—40 Quarts		
	Cans	Quart bottles	Pint bottles	Cans	Quart bottles	Pint bottles
Zone No. 1: 1 to 40 miles...	$0.23	$0.31	$0.35	$0.41	$0.49	$0.53
Zone No. 2: 41 to 100 miles.	.26	.34	.38	.44	.52	.56
Zone No. 3: 101 to 190 miles.	.29	.37	.41	.47	.55	.59
Zone No. 4 over 190 miles...	.32	.40	.44	.50	.58	.62

The rates continued to be raised and complaints increased.

STATEMENT SHOWING INCREASE IN FREIGHT RATES ON MILK SINCE 1902
Zone No. 1—1 to 40 Miles

	Milk in Cans, Per Can		Milk in Quart Bottles, Per 40 Cans	
	Carload shipments	Less than carload shipments	Carload shipments	Less than carload shipments
Aug. 1, 1902—Sept. 30, 1903.........	.184	.325	.248	.31
Oct. 1, 1903—Sept. 30, 1909.........	.184	.23	.258	.322
Oct. 1, 1909—Feb. 22, 1915.........	.201	.23	.282	.322
Feb. 23, 1915—to 1917.............	.211	.242	.296	.338
Per cent of increase since 1902......	9.2%	2.9%	15.3%	.9%

Zone No. 2—41 to 100 Miles

	Milk in Cans, Per Can		Milk in Quart Bottles, Per 40 Cans	
	Carload shipments	Less than carload shipments	Carload shipments	Less than carload shipments
Aug. 1, 1902—Sept. 30, 1903.........	.208	.26	.272	.34
Oct. 1, 1903—Sept. 30, 1909.........	.208	.26	.291	.364
Oct. 1, 1909—Feb. 22, 1915.........	.228	.26	.319	.364
Feb. 23, 1915—to 1917.............	.239	.273	.334	.382
Per cent of increase since 1902......	14.9%	5.%	22.4%	12.4%

MILK FREIGHT RATES

Zone No. 3—101 to 190 Miles

	Milk in Cans, Per Can		Milk in Quart Bottles, Per 40 Cans	
	Carload shipments	Less than carload shipments	Carload shipments	Less than carload shipments
Aug. 1, 1902—Sept. 30, 1903.........	.232	.29	.296	.37
Oct. 1, 1903—Sept. 30, 1909.........	.232	.29	.325	.406
Oct. 1, 1909—Feb. 22, 1915.........	.254	.29	.355	.406
Feb. 23, 1915—to 1917	.267	.305	.373	.426
Per cent of increase since 1902......	15.%	5.2%	26.%	15.1%

Zone No. 4—over 190 Miles

	Milk in Cans, Per Can		Milk in Quart Bottles, Per 40 Cans	
	Carload shipments	Less than carload shipments	Carload shipments	Less than carload shipments
Aug. 1, 1902—Sept. 30, 1903.........	.256	.32	.32	.40
Oct. 1, 1903—Sept. 30, 1909.........	.256	.32	.358	.448
Oct. 1, 1909—Feb. 22, 1915.........	.28	.32	.392	.448
Feb. 23, 1915—to 1917.............	.294	.336	.412	.47
Per cent of increase since 1902......	14.8%	5.%	28.7%	17.5%

In 1916 the Interstate Commerce Commission made a study of the subject and on October 1, 1917 prescribed a mileage rate in 10 milk blocks extended to 630 miles. The rate then was increased 25 per cent in 1918. It was increased a further 20 per cent in 1920. The less-than-carload rate for a 40-quart can in the 10 mile zone or under was 23.5 cents; for 90 to 100 miles, 35.5 cents; 200 miles, 45 cents; 599 miles, 65½ cents. Car rates were 87½ per cent of the less-than-carload rates. The shipper furnished the ice.

Rates were reduced in 1933. The table on page 54 shows the rates at that time on for less-than-carload lots:

The development of hard-surface State roads and the perfection of the auto tank truck resulted in competition with the railroads for the transportation of milk. Tank cars were also utilized by the railroads. Since 1934 the tank cars have very greatly increased the volume of milk handled, and the cost of transporting milk from the 201-210 mile zone has been reduced to something less than 22 cents per cwt., and in some instances as low as 19 cents per cwt.

Miles	Freight Rate per 40 qt. can Class 1 Milk	Freight Rate per 100 lbs. Class 1 Milk	Miles	Freight Rate per 40 qt. can Class 1 Milk	Freight Rate per 100 lbs. Class 1 Milk
10 or under	$.20	$.235	201-210	$.385	$.355
11- 20	.21	.245	211-220	.40	.47
21- 25	.225	.265	221-225	.405	.475
26- 30	.225	.265	226-230	.405	.475
31- 40	.235	.275	231-240	.41	.48
41- 50	.25	.295	241-250	.41	.48
51- 60	.26	.305	251-260	.425	.50
61- 70	.27	.32	261-270	.43	.505
71- 75	.28	.33	271-275	.435	.51
76- 80	.28	.33	276-280	.435	.51
81- 90	.29	.34	281-290	.44	.52
91-100	.30	.355	291-300	.45	.53
101-110	.305	.36	201-310	.455	.535
111-120	.315	.37	311-320	.46	.54
121-125	.325	.38	321-325	.465	.545
126-130	.325	.38	326-330	.465	.545
131-140	.335	.395	331-340	.475	.56
141-150	.34	.40	341-350	.48	.565
151-160	.355	.42	351-360	.485	.57
161-170	.355	.42	361-370	.49	.575
171-175	.365	.43	371-375	.495	.58
176-180	.365	.43	376-380	.495	.58
181-190	.375	.44	381-390	.50	.59
191-200	.385	.455	391-400	.505	.595

CHAPTER VIII

STUDY OF DISTRIBUTION ADVOCATED

Farm Co-operation Proposed.

When the State Department of Agriculture took over the management of the State Fair and the Farm Institutes, the New York State Agricultural Society functions became dormant and remained so for some years. During the late Fall of 1909 Raymond A. Pearson, then Commissioner of Agriculture, called a conference of its members to discuss the possibility of a renewal of its activities. The conference brought out the facts that the once comprehensive agricultural services of this Society had been completely assumed by special State organizations devoted to the specific branches of agriculture such as horticulture, floriculture, livestock breeding and dairying. This was the case even before it had limited its activities to conducting the State Fair and the Farm Institutes. Because of this situation there was considerable feeling in the conference that there was no useful place left for the old society.

While admitting the facts in the main, I reminded the conference that the existing organizations were designed for the improvement of varieties and breeds and increased production. No organization existed for the purpose of developing a more efficient and economic system of distribution. Some of my good friends said in so many words that this was a worthy purpose but that it was "ideal and not practical." After the possibilities were pretty fully explored, however, it was decided to appoint several committees to make recommendations and report at a general meeting to be held on the dates set in the by-laws of the Association. This was the third Wednesday in January, 1910.

I was appointed chairman of a committee, consisting of Augustus Denniston of Orange County and myself, on "Plans and

Scope." It fell to me to write the report in which my associate fully and enthusiastically concurred without a single word of dissent. While it included several worthy purposes advanced at Commissioner Pearson's conference, I stressed particularly the need of an organized marketing system for the products of the farm and pledged the New York State Agricultural Society to advocate that objective. The policy was endorsed by several speakers and adopted with enthusiasm at the General Convention on January 18-20, 1910 by a large body which fairly represented the best farm thought and practice of the State.

For a decade following, farm co-operative marketing was the outstanding subject discussed at the annual meeting of the society. The proceedings were printed in pamphlet form and are on file in the Department of Agriculture. Some years later I received a letter from Hon. Berne A. Pyrke, then Commissioner of Agriculture, saying that he had just concluded a study of the reports of the New York State Agricultural Society during the period referred to above, and he was writing to express his commendation of it. He said that he believed the farm marketing problem would have to be worked out under the fundamental principles and ideals which I had developed and outlined in these discussions.

At the beginning there were many sympathetic members in a mood to be shown. We seldom mustered a dozen outspoken enthusiasts for farm co-operation. Among these were Ezra Tuttle, Fred Boshart, Fred Sessions, George Sisson, Charles White and Patrick Barry. It was encouraged by Commissioner Pearson and later by Commissioner Huson. By 1912 the "high cost of living" had helped create interest in the subject and consumers, as well as producers, had become interested in the spread between the prices farmers received for food and what the consumers paid.

Action by State Agricultural Society.

Encouraged by this general interest the New York Agricultural Society called a two-day conference of delegates representing producers' and consumers' organizations of the whole State, with a view to the cheaper distribution of food products through co-operative methods. The conference was called by Mr. Sisson,

STUDY OF DISTRIBUTION ADVOCATED

then president of the society. It was held in New York City on April 19 and 20, 1912. That conference represented twenty-two organizations of producers and consumers, all the active farm organizations and practically all the social and economic consumers' organizations, and industrial interests.

The meeting was well organized and ably conducted. There was no dispute about the facts, nor of the need of a reform of the distributive system, but no plan suggested was generally approved. My position was that the convention represented too many emergency interests. The temporary emotions would soon subside. The farm problem, on the contrary, was permanent. As long as city people had food to eat, farmers would have to produce it, and no one could be expected to look after farmers' welfare in the enterprise but themselves.

Near the close of the two-day discussion it was evident that nothing definite had been accomplished, but just before adjournment a motion was unanimously approved to create a State Standing Committee on Co-operation to work out a plan for the distribution of farm products. It provided that I should be chairman of the proposed committee with authority to select the associate members.

I took this assignment seriously. I invited about one hundred or more outstanding farmers, widely distributed over the State, to join me in the committee. They accepted and all were helpful; many enthusiastically so. It marked the beginning of organized farm co-operation in New York State. Up to that time farmers had organized under the business laws as stock corporations. The committee therefore secured the enactment of a special farm co-operative law authorizing incorporation of the membership type of associations to conduct the business of the membership. It was a bit crude, and experience soon showed that it needed revision, but it was a beginning. We were blazing a trail. We had been working vainly for some years to get a State law to regulate the commission sale trade in the New York markets. The committee focused attention on the subject, and succeeded in inducing the Legislature to pass its bill the next year in the face of powerful trade opposition.

During the summer of 1912 I went to Europe to study the marketing systems abroad. There was much in the management of the Halles Centrales, the great food and produce market in Paris, that could be profitably adapted to similar markets here. It is a public state market occupying some twenty-odd acres in the heart of the city. It is completely under public control. Its management is subject to definite mandatory laws.

In 1912 the New York farmer was getting about thirty-five cents and the distributor sixty-five cents of the consumer's dollar. As near as I could determine, this allocation of the consumer's dollar was more than reversed in Paris. In many cases the farmer got eighty cents of the consumer's dollar and there was no gouging. The farmer got a check showing the weight of the sale from an official automatic scale and a record of the price with the name and address of the buyer, together with the commission charge when sold on consignment. In America there are so many vested interests in the distribution of farm products that the organization of similar markets under appropriate laws has been so far defeated.

While in Europe I also studied the foreign "farm credit" system which had begun to arouse much favorable discussion in this country. I did not find any system of marketing or of farm credits in Europe that I thought would serve our purpose here, but I did find features in their systems which I believed could be profitably adapted to our conditions and traditions. As to commercial farm credits, their system in Europe was too petty to interest farmers in America, or to serve them efficiently if attempted.

State Land Bank.

For commercial purposes our local banks afforded better promise. For the permanent financing of farm mortgages we had no system at all. New York State, however, had a splendid system of co-operative savings and loan associations to finance the mortgages of their members, but the members' deposits did not furnish funds enough to meet the demands for home mortgage loans.

In some locations, however, the local savings and loan associa-

STUDY OF DISTRIBUTION ADVOCATED

tions found the farm mortgage an attractive investment for its surplus funds.

A committee representing the savings and loan associations and our co-operative committee conceived the idea of affiliating these associations in a "State Land Bank" and using their combined credit for the sale of long-term bonds to provide funds for mortgage loans for both residential homes and farm property. The joint committees developed a charter for the Land Bank. The State Banking Department co-operated. Governor Glynn appointed Henry Burden and myself as an auxiliary committee to the State Banking Commission, then revising the State banking laws, that we might be in position to further the interests of the proposed Land Bank. The charter was approved and the Land Bank of the State of New York was organized that year,—1914. It is an ideal co-operative system for financing mortgages and is successful in the field of home mortgage loans. Its farm service was never developed because of the organization of the Federal Farm Loan Banks with advances of government money.

When I was in Europe I made the acquaintance of the Ambassador to France, Myron T. Herrick, which later developed into a delightful friendship. Mr. Herrick became interested in a proposal for a system of Federal farm mortgage credit. At his request I became a member of his committee for the purpose under the auspices of the United States Chamber of Commerce. In the final outcome neither of us got all we wanted. We desired a more complete farm control than the system that prevailed. Mr. Herrick was never fully reconciled. The set-up of the Farm Loan Associations was a further objection, but as a whole the Federal Land Banks with government advances of money and guarantee of bonds have formed a more comprehensive system than any private institution could be. It followed our State plan in many essentials, but it is not co-operative. Getting its original millions in funds direct from the National Treasury, it replaced our New York system, so far as farmers were concerned, before the Land Bank had time to develop.

My excuse for this digression from milk is that many dairy farmers were and are interested in mortgage loans.

Report of State Standing Committee.

At a meeting of our State Standing Committee in the Fall of 1913, reports of sub-committees revealed the farm marketing conditions clearly and frankly before the members. The market was entirely under the control and domination of middlemen. They had multiplied themselves needlessly in the trade. They fixed the price to suit themselves. In addition they made charges against the farmer for freight and demurrage they never paid, for cartage on goods that were sold on the wharfs, and for damaged goods that they sold for perfect quality. They sold and delivered perfectly good products to scalpers and sent the farmers certificates that the shipment was condemned by the health authorities. They sold produce at the market price and returned the farmers a lower price. They rented desk room in a produce building, printed pictures of the front as if they owned the building, and flooded producing districts with circulars and letters quoting prices above the market. They never made any returns. When producers became insistent they disappeared as suddenly as they had first appeared. No one believed that the merchants who rented the desk room let the crooks get away with the plunder without a share in the larceny. Milk dealers fixed the price of milk in an illegal and fraudulent exchange and used practical tricks to cheat the farmer out of a part and often the whole of his shipment.

The worst of it all was that these middlemen were protected in the city and State government, not by direct action of either government, but by individual members in them and by abundant funds to use when and where needed. The system was built up by these influences. It was so maintained. It could not, I believe, be created or kept up without these political influences.

The national political campaign of 1928 developed a demand for a marketing system for the distribution of farm products. The following year I was invited to appear before the United States Senate Committee. I interested the Committee in a plan for a series of wholesale farm markets in which co-operatives would have space for the sale of special crops, for both domestic and export trade, but the organized pressure of the central west prevailed. The half a billion dollars and more wasted in the Federal Farm Board would have given us a national chain of markets that would be of priceless value to the producers and consumers of the whole country.

CHAPTER IX

THE DEPARTMENT OF MARKETS

The Plan Approved.

Quite in dismay by the revelations a member of the committee asked, "What are you going to do? Give up?"

"No," I said, "I have no thought of giving up. These people have a system under cover. They operate it by touching secret wires under the table. I propose that we get right out in plain view on top of the table and build a system in plain view of all the people, producer and dealer and consumer alike."

"How are you going about it?" a member asked.

My reply was, "I propose that this committee prepare a bill creating a State Department of Markets, as independent and free to act as any other department of the State, and ask the Legislature to pass it."

"How long do you suppose it will take to get it through?" asked the previous inquirer.

"I don't know," I replied. "It may take one year. It may take six, but we will be working for a definite objective. This alone will tend to check the abuses and save us many times the cost. We will gain the open-support of sympathetic friends in the Legislature who keep silence now because we have given them no definite issue." The committee approved the suggestion for a State Department of Markets.

Governor Glynn.

One day in the early part of January, 1914, I received a request from Governor Glynn to come to Albany. He had heard that I had gone to Europe to study "farm credit" and he wanted to know what I had learned. After talking about it for a time, he concluded that he might have to drop the idea. He seemed a bit disappointed. As I was about to leave, he remarked,

"I was not farm bred but I was brought up in a small village in Columbia County, and I know farm people. They have had more than their share of hardships. I have seen laws passed that promised to help them, but none of them seemed to have done the individual farmer any good; so I have always thought that if I ever got into a position to do so, I would do something that would be a real worthwhile benefit to them."

Up to that time I had assumed that he was looking merely for something that would make a political appeal, but in this there was no mistaking the sincerity and disinterestedness of a personal sentiment. It impressed me, and in the intimate associations that I had with him during the remainder of his administration and to the time of his death I found in public and in private that same sincere sentiment for the farm directing him in all problems affecting agriculture. I had never met him before. It was the beginning of our acquaintance and of our friendship. I replied to his confidence,

"Governor, since you want to do something really worthwhile for farmers, I can give you a proposition that is worth to them a thousand to one of co-operative credit."

Speaker Sweet.

Our committee had already made a rough draft of a bill to create the new department. Governor Glynn became enthusiastic over the proposition and had the Bill Drafting Bureau put the bill in shape for introduction. The Governor was an independent Democrat. Democrats had a majority in the Senate and with the Governor's support we felt sure of the Senate. The Republicans had a majority in the Assembly. One morning I went to see Speaker Thaddeus Sweet in his room at the Capitol. I told him briefly the purpose of the bill and the need of it. He had heard of it. He said very frankly but not unkindly in tone, "No, sir. As I understand it, I do not like the source of this bill and the Assembly would not stand for it."

I was inexperienced in the task I had undertaken, but I briefly explained to the Speaker that our committee had been selected without consideration of their political affiliations, that some of them were his personal friends and probably many more Republicans than Democrats in its membership, and that to my personal knowledge the farmers of the State wanted the Market Department established. Then I made this appeal,

"Assuming, Mr. Sweet, that the farmers of the State want this legislation, would you like to have it published in newspapers and in the agricultural press of the State that a Democratic Governor and a Democratic Senate, or if you please, a Tammany Hall Senate, are willing to give it to them, but the Republican majority of the Assembly that farmers made possible by their votes, refused to help them?"

It was already past the time to open the Assembly for the day's session. Speaker Sweet said so as a reply to my appeal but invited me back at three o'clock that afternoon.

In the afternoon he asked some pertinent and proper questions. "How did this committee originate?" I told him.

"Were all the members farmers?"

"Yes."

"Who paid the expenses of your trip to Europe?"

"I paid my own expense."

"Who paid the expenses of the State Standing Committee on Co-operation?"

"The members paid their own expenses. I paid any expense I contracted in the way of postage, correspondence, etc."

"If this bill passes, who is to head it?"

"I do not know. We have never even mentioned that part of the subject."

"Do you expect to head it?"

"Positively and definitely no."

At this interview Speaker Sweet told me he had come to consider our bill more favorably. He wished to consult some of his associates in the Assembly during the evening and he would write me as I was returning to New York that night. His letter advised me that he approved the bill and that himself and his friends would support it in the Assembly. He did so loyally. He was one of my most helpful friends during the years that he remained a member of the Legislature.

The Department of Foods and Markets.

The bill passed both houses of the Legislature and was signed by the Governor. It created the Department of Foods and Markets, and provided for one commissioner to be appointed by the Governor for a term of six years at a salary of $6,000 a year. The commission had authority to investigate all phases of marketing and prices of foods and farm products. It was charged with the duty of helping farmers in the organization of co-operative associations for the marketing of farm products, to assist farmers through such organizations, and create an efficient and economic system for the distribution of farm products. It was, I believe, the first State department in America devoted to the marketing of farm products.

The Commissionership.

As soon as the bill became law, a scramble began for the Commissionership. The New York City produce dealers had a candidate. The milk distributors had suggested a man for the place. Altogether there were fifty applications. Our committee had filed an application. Governor Glynn delayed the appointment until November. He told me he had never had so much pressure for an appointment. Opposition had so particularly focused against our candidate that he just could not name him for the place. He had sent the committee at different times the names of three different applicants. After investigation we felt one was impossible, and the other two, while unobjectionable gentlemen, did not have the qualifications we felt to be essential for the place.

He then appealed to me to take the place as a way out for him. He knew of my devotion to the publication of *The Rural New-Yorker,* also that I had not fully recovered from the effects of a case of double pneumonia contracted in Switzerland at the time of my European trip, but he argued that I was the person who should organize the Department and that, this accomplished, and its work initiated, I could retire. He would, he said, explain the situation to Speaker Sweet because of my assurance to him that I was not to have the place.

The Speaker was advised of the Governor's difficulty. In the meantime a mutual confidence and friendship had developed between Mr. Sweet and myself and he approved the appointment. I, however, persisted in my refusal to take the place for two weeks or more, but at the insistence of members of the Committee I finally accepted the appointment in the early part of December, 1914.

CHAPTER X

MILK CAMPAIGN STARTED

Department of Foods and Markets.

One of the first acts of my administration of the department was to call a conference of dairy farmers to consider what we should do for milk producers. The sentiment was that, if a major number of dairy farmers would consent to place the sale of their milk in the control of the department, a satisfactory price could probably be negotiated. If not, the department should proceed to develop a system of distribution. To that end we realized that it would be necessary for the consenting farmers to create an association, because the department was not authorized to do business, but it was authorized to help farmers organize and to assist the organization in marketing all farm products.

Dairymen had been discouraged by the failure of several attempts to organize since 1882. We knew that the need existed and that there was some sentiment for a new trial, but nothing could be done without the co-operation of the rank and file. Of course, it fell to me to find out how far they would go with a new proposition. I began to interview dairymen by letters, at meetings and personally. I found most encouraging sentiment everywhere but in numbers there were not enough.

I felt the demand should be more general. During the Summer of 1915 I therefore distributed in the milk-producing counties a petition which farmers might sign and return if they wished the Department to go ahead with the proposition. I also sent the blanks to local papers in dairy areas. It received considerable publicity. The Poughkeepsie Sunday *Courier* published the petition, preceded by editorial approval as follows:

Milk Producers: Sign This Petition!

"Farmers and milk producers have an opportunity now to secure an investigation by State experts to show to the public the exact figures of the cost of milk production and distribution."

"This will offer convincing proof of whether the dairymen or the distributor is responsible for the advance in milk prices and whether the farmer is getting his share of the advance."

"No dairyman fears what the figures will show. Every branch of the farm business except that of the dairy shows largely increased profits during the past several years; but many dairies are now conducted at a loss or with profits so small as to be almost negligible, while the public pays high prices for milk. Poughkeepsie dealers pay higher prices than the trust does for milk, and the cost to the consumer is no higher in this city than in the milk trust's territory."

"The experts of the State Department of Foods and Markets will undertake this investigation with all the legal resources and power of the State to back them up. If the farmers and dairymen in this section will sign the petition printed herewith, cut it out and mail to the *Courier*, we will forward it to F. H. Lacy, Manager of the Dutchess County Farm Bureau, who is co-operating with the State Department of Foods and Markets."

"PETITION

"To John J. Dillon
Commissioner New York State Department of Foods and Markets
Nos. 202-204 Franklin Street
New York City.

"We, the undersigned farmers respectfully petition you to make a thorough investigation of the milk business in the State of New York with a view of ascertaining if it is not possible to bring about changes in the business, whereby a price at least one cent per quart in excess of the average price of milk for the past three years, shall be paid to the farmers of New York State for the next three years. We would also like to have the largest possible distribution of milk in all the large consuming centers and, if possible, a lower price to the consumer than has been charged in the past three years.

Name	Post Office Address	No. of Cows"
....	....	....

Farmers promptly realized the influence and power that a State department would bring to their cause. They signed the petitions and returned them in large numbers and with cheering comment. All of them approved the proposal to increase the price of milk one cent a quart. It meant an increase of thirty per cent in the highest price paid and more than thirty-three and one-third per cent in the average price then being paid.

The general interest of dairymen exceeded anything I had seen in forty years' experience. It was mid-Summer before this sentiment had fully manifested itself, so was too late to attempt a sale that October. April was the next contract date, and with a flush of milk in April, that was no time for farmers to open a milk price campaign. There was nothing to do but prepare for October 1, 1916.

So many failures had followed attempts to organize associations with initiation fees, dues and assessments, I was determined not to ask for money or pledges of any kind except that farmers verbally promise to make the department the sole agency for the sale of their milk collectively and to refer dealers who approached them to the State Department of Foods and Markets.

I thought I knew farmers well enough to know that if they made a promise they would keep it, and that if I did my part they would do theirs. They did. The one thing they persisted in advising was, as they expressed it, not to attempt to "revive any corpses."

"Have nothing to do with the 'Five States' or the Dairymen's League," they advised. "Begin clean and build up an entirely new organization." The experience of the Five States Association has already been told.

The Dairymen's League was the last to attempt organization. It had started back in 1904 and was incorporated in 1907, under New Jersey laws with a good business stock company charter. The stock was sold to dairy farmers by paid canvassers on the basis of twenty-five cents per cow. The purchaser of the stock became a member of the organization. Proxy votes were allowed. In about ten years some $50,000 had been collected and spent to sign up about 13,000 members.

The organization had sold no milk and made no accounting of

the money collected to stockholders. I felt, however, that the failure was not due to any intentional misuse of the funds, but to inexperience and lack of understanding and energy. I personally knew only one of the official group, F. H. Thompson of Holland Patent, and one director, Harry W. Culver of Amenia. I had confidence in the integrity of both men and never lost it. I liked the name and hoped to reconcile the objectors to it. I therefore held the matter in abeyance but continued the discussion of milk in speech and through the mail and the press. Among various bulletins, I issued the following statement about the first of December, 1915. It was published in *The Rural New-Yorker* on December 4th:

The Big Milk Question

Farmers from all parts of the State have appealed to the Department of Foods and Markets to help find a profitable market for their milk. The complaint is that everything used in the production of milk has increased in cost during recent years, while the price of milk has remained practically the same or in some cases even less. In addition to the increased cost of material, labor and capital, a grade of milk has been exacted that costs more to produce. Many of the best milk farmers of the State are going out of the milk business, and those who remain are not increasing their production. This is creating a short supply except in a comparatively short time in the Spring when there is a surplus for a few weeks only. If the present progress of elimination continues, the shortage must become chronic and the price to the consumer will be again advanced without any corresponding increase to the producer.

And so here we are again up against the old vexed milk problem. The solution of it has been attempted many times by farmers' organizations and by legal processes leveled against the dealers' exchanges. The legal processes have been technically successful without helping the situation, and the organizations of producers for one reason or another have so far failed, either through a want of cohesion of their membership or through treachery in their ranks or weakness in their leaders. For a time these experiences discouraged the producers and they returned to the condition where producers contracted individually and independently of one another with dealers knitted together in one of the closest compacts of the business world of the day. It is only the clever business sense of this monopoly that keeps the price from going lower. To reduce it further would be to cut off the supply entirely, and this they have no wish to do.

Can the Department hope to adjust differences of such long standing? Can it win in a field where there have been so many failures? If experience were to be accepted as conclusive, the present task would seem hopeless, and yet the Department has consented to tackle the job. Will it win? Yes, we make this prediction on the strength of a sincere leadership, and a loyal body of producers. The American farmer has been accused of failure to work in double harness. He has been called a quitter. He has been accused of failure to accept measures proposed for his own interest and charged with ingratitude towards those who befriended him. If these things were true the movement we are discussing would fail. But these things are not true. The farmer like other human agents makes mistakes at times and all mistakes are costly. He has been known to punish a friend to support a principle. Most of the indictments against him have been drawn by schemers who had selfish interests to promote. The farmer hesitates where there is a suspicion of insincerity. He is not so likely to surrender in face of danger as he is prone to retire in disgust where he detects treachery in the leaders or in the ranks. Give him a leader he can trust and a cause that is just and he neither hesitates nor surrenders. We feel some justification in this assertion because we have led him in battle and tried his mettle. Where a moral question was involved and the principle made clear the American farmer never hesitated and never faltered. He fought.

We predict success in this undertaking because we have trust in the leadership, faith in the soldiery and confidence in the cause. It will require brains and work and sacrifice and loyalty. All this will be given it; and the cause will win.

Proposed Economies.

As an auxiliary to the proposed system the plan contemplated a milk depot in the City of New York, where small dealers could be supplied and an annexed creamery equipped to pasteurize and to work up any surplus or over-ripe milk that might remain on hand. I proposed a milk depot to operate this plant by a co-operative association of producers, with books wide open to demonstrate the economic cost of distribution and furnish a basis for negotiating prices with milk dealers. I was confident that through economies farmers could negotiate a price on the cost of production and a reasonable profit, pay distributors the cost of their service with a profit and yet reduce the cost to consumers. This was expected to increase consumption and production, leading to further economies and lower costs.

The Wicks Committee.

I knew the big dealers would fight this program to the death. I knew it would break the price monopoly they had held, with only one short exception, for forty years. The big dealers knew it as well as I did. When it came to negotiating prices and terms farmers would be in a position to protect themselves. Farmers, however, did not want to disturb the distributors, but they did want a price to cover cost of production and a reasonable profit. This organized system would enable them to demand and get it. The big dealers would get a smaller profit but they would get all they earned and their business would be on a stable basis with all concerned fairly protected.

This proposition brought enthusiastic and practically universal approval from producers. City store keepers approved it. During the following year the butchers of the Bronx visited me several times in reference to it and finally agreed to sell bottled milk and cream from all their stores at our price, and to begin just as soon as I could arrange to supply the milk as their increased trade demanded. They had refrigeration and delivery wagons. They agreed to deliver for one cent a quart because milk is in universal demand and the delivery would help their meat trade. The sequence of this plan will come up later on.

I prepared a bill to facilitate these purposes and asked Senator Charles W. Wicks to introduce it. He seemed greatly pleased to have the privilege of doing so, representing as he did, a dairy district in central New York. While I was in Albany, Assemblyman H. E. Machold, an influential member of the Legislature, volunteered to introduce it. I should have gone to him first, but I had spoken to Senator Wicks, who had promised to introduce the bill the next day.

Two days later Senator Wicks called me on the telephone and said there was some objection to the bill. I at once started for Albany and sought Senator Wicks and soon discovered that the opposition came from Senator Elon Brown, temporary president of the Senate. Senator Wicks and I went to his office. We argued the matter back and forth for some time. At one time consider-

MILK CAMPAIGN STARTED

able heat was generated by both of us. Senator Brown had denied some fundamental rights of farmers which I had advanced. I replied,

"Senator, I could go into your dairy district and defeat you for re-election on that issue."

"Go and do it," he hotly retorted. I assured him I had no wish to do so. Then he said in a friendly tone,

"No use of our disputing over it. What you need is an investigation of the milk business. That we will give you."

I did not want any more investigations. I reminded him that Attorney General O'Malley had recently made an exhaustive investigation and nothing had changed since except for the lower price of milk, and we all knew what was the matter with the milk business. I could give him the history of it for fifty years back.

I saw, however, that I could not break down his resistance and knew I could not win against the Temporary President of the Senate. I thought an honest investigation might strengthen our position in the Legislature in the next session, and help focus farm sentiment into action. On this theory I said that if he would appoint Senator Wicks as chairman and begin the investigation as soon as the Legislature adjourned so that we could have the effect of it in before the time for negotiating prices for October 1st delivery, I would consent. He readily agreed. At his request I prepared and sent him a memorandum for the resolution which he used much condensed in the preamble of the resolution, creating the Committee. On April 7, 1916 Senator Wicks sent me a copy of Governor Whitman's letter expressing hearty approval and support of the resolution.

The Committee did not get under way on time but it did some good work up-State during the Summer. I think its revelations helped stiffen farmers' determination for the October fight.

The State's Plan.

This was followed during the same month by the outline of a definite plan and procedure for the sale and distribution of milk in New York City and State. The plan was to organize the producers supplying each and every local plant into a local association with sovereign authority in its local affairs and with exclusive control in the farm membership, limited to local producers. Facilities were provided to safeguard farmers collectively in control of the association, its management, its policies and its business. This included one vote by ballot for each and every producer. The raising of money, the size of the budget, the disbursement of money and the salaries to be paid were all to be authorized by ballot votes. It was provided that all officers be elected annually by ballot vote and that officers be rotated so that no officer could succeed himself for more than one term, thus limiting him to two successive years.

It was provided that all of these local associations be affiliated in one State body which would be under the control of the individual farmers. The only restriction on the producer and the local association would be the one voluntary self-denial of their right to negotiate the sale and price of their milk. The one State-wide body was charged with this responsibility in form, but it was under the control of the associations, and every farmer was given a voice in fixing the price and the terms of sale.

It was anticipated, and a part of the policy, that many of the local associations would build plants to handle the milk of their members, and possibly process or perhaps manufacture milk.

It was also anticipated that, when farmers were so organized and equipped, milk dealers would co-operate with the farmers in negotiating a fair and equitable price. It was admitted that the dealers served a useful purpose for which they were entitled to fair pay. Farmers needed the service and were willing to pay for it. But dealers did not have a legitimate, exclusive right arbitrarily to fix the price of milk for the men who produced it. While farmers needed the dealers' service, the dealers needed the farmers' milk and could not exist without it. Farmers were simply adopting a business method to protect their own heretofore neglected interests.

CHAPTER XI

THE DAIRYMEN'S LEAGUE, INC. IN 1915

Secretary Manning's Testimony.

A record made by the Wicks Committee in the following year really belongs at this place in my narrative. On September 25, 1916, Albert Manning, Secretary of the Dairymen's League, responded to a subpoena and testified before the Wicks Committee in the city of Utica.

At the beginning George W. Ward, the counsel, made a statement in which he said, substantially, that sometimes a dairyman appears who is an enthusiastic member of the League. "Sometimes," he said, "we meet dairymen who are very skeptical, to say the least, of the uses and purposes and methods of the Dairymen's League. We meet dairymen who say 'What do you people find out about the Dairymen's League? Their representatives have come here and asked us to contribute money. What do they do with this money?' We thought that a statement that dairymen and the committee and the Legislature could understand would be useful."

Mr. Manning testified that stock representing 190,000 cows had been sold and that would mean about $47,000. Mr. Ward stated that some dairymen had come there and said that non-assessable stock was bought and paid for and that assessments had been levied on the original amount of stock the current year. Mr. Manning testified that the by-laws authorized the levy. The money collected for stock and for assessments had been paid out to canvassers, for expenses of directors attending meetings, for incidentals and for salaries of officers.

Chairman Wicks asked, "Then most of that money was really lost, was it not?"

Mr. Manning replied, "It didn't do very much good—didn't accomplish very much."

Assemblyman Law asked, "Up to the first of January you had received about $50,000. What benefit did the dairymen receive for that $50,000?"

Mr. Manning said that the benefit might be more indirect. "When the organizers could not get enough stockholders to pay their expenses, we stopped them," he said; and that there was "no active organization that year."

When asked the reason for it he said, "Well, the organizers were not able to get enough new names to pay their expenses, and as soon as they had covered the territory to that extent, we stopped the work."

The officers voted an assessment of twenty-five cents on the stockholders in the early part of 1915, mailed a letter with the call, and collected from $3,000 to $4,000. The returns "exceeded expectations" he said and further that the money was used up in the same way as that received from the sale of stock. Asked the position of the organization on the matter of strikes, Mr. Manning replied that it had told the dealers the price wanted, and he didn't know whether they would pay it or not, but the League was not committed to any definite line of action.

R. D. Cooper's Correspondence.

Under date of June 5, 1916 I received a letter from R. D. Cooper, Little Falls, N. Y., who was chairman of the executive committee of the League, consisting of R. D. Cooper, F. H. Thompson and Frank Sherman. In this Mr. Cooper said a meeting of some co-operative plants was held at Utica on May 27th. There were not many plants represented but enough to start a federation of farm-owned plants. A Mr. Myer of the Markets and Rural Organizations of Washington was present and had taken "our" proposed articles of agreement for study, stating that he would return within ten days. "Mr. Thompson and I," said Mr. Cooper, "believe it advisable to consult with you [Mr. Dillon] as to linking this proposed organization in with the plan with which you are working."

The letter also stated that the directors had met in Albany on June 1st and directed its executive committee "to adopt and carry out such plan and policy as will result in procuring such price for milk as will be approved by the directors." The last paragraph read:

"It looks very much as if a strike will be called, unless other means can be employed to secure the necessary raise. This action taken at the directors' meeting was not recommended by the executive committee, but the board of directors are of the opinion that the co-operative system upon which we are working is too slow. The demand has been made and therefore the board of directors must do something to justify their existence. Personally, I am in hopes that perhaps some way can be worked out whereby the desired raise may be secured with the aid of the services of your office. Can you think of any possible way by which milk can be marketed under your supervision without being handled through local co-operative plants? Do you believe it would be possible to secure a raise in the price of milk by taking it up with the representatives of the New York dealers, such as Borden's and other large Companies?"

R. D. COOPER.

In my reply on June 7th, I told Mr. Cooper briefly that in my judgment it would be a mistake to undertake a strike. I thought a series of strikes would not solve dairymen's problems and that "we might just as well face the situation and prepare to handle it in a way that will bring permanent relief." I believed that an appeal to the milk dealers would not be successful until we got into a position to demand what we wanted. They knew our weakness as we stood. I related my purpose as I had previously outlined. It was, I reminded him, a peaceful business way to put farmers in a position to market their milk through dealers if they would pay a reasonable price for it, and to market it otherwise if they still refused, and that it could be done for less than the cost of a strike.

I received a prompt reply in which Mr. Cooper approved my plan and would like to see it tried out. He continued:

"Personally, I talked and advised against the strike at our meeting in Albany, June 1. As you know, Mr. Thompson is also opposed to that method of doing business, and Mr. Sherman another member of the executive committee stated that in the end the co-operative plant scheme was the only one. Most of the other directors present were hot for a strike and we told them that if they were bound to have one, that we would do the best we could. I sincerely hope that we can work some scheme other than employing a strike method. I have already sent you extracts from the minutes of that meeting which does not state that we are going to call a strike but the demand seems to be for one.

In the course of the next ten days or two weeks I would like to have a conference with you in regard to these matters, but prefer to have it after we have gone over the matter with Mr. Brand or his representatives."

In my reply I approved his suggestion for a conference and wrote further:

"I am not so hopeful of the help from Washington. Mr. Brand is a good fellow, and a capable man, but up to the present time these Department men have been so cautious and so conservative about everything undertaken along this line that I have gotten into the way of not looking for anything aggressive from any of them, and our proposition demands new precedents and some thing entirely out of the ordinary procedure. Of late there has been a disposition in some departments and institutions to recognize the need of something more than research and investigation and conservation in the handling of these business problems. The Department at Washington would be a great force and a great help if it would take a hold of the work with us in an aggressive way. If, on the other hand, they are disposed to stand on legal technicalities and conservative methods they will be a hindrance rather than a help to us. Let us see what they will do, but let us not delay our work too much waiting on their motion.

"If they will go ahead and lead in the fight, you and I could well afford to let them have the leadership and we could work behind them in the ranks. For myself, I am willing to do so. The position of the National Department of Agriculture leading this move would be a strong card, and I would be glad to subordinate this Department and myself personally to their leadership, but frankly, from the experience of the past, I do not think we would be justified in assuming that the Federal Department will take this leadership. I am rather inclined to think that they will be more disposed to put on the brakes than to press down on the accelerators. However we can wait and see, and the only justification of this discussion is that we do not delay too long waiting for them in the development of our plans."

I do not know that the Washington lead went any further, but

I felt satisfied that in this case Mr. Brand could not have found a situation to justify his immediate participation.

H. J. Kershaw's Distrust.

The helpless position of the Dairymen's League was a matter of general information in 1914, 1915 and 1916 up to the time that announcement was made that the milk would be sold through the State Department of Foods and Markets, which farmers themselves had just caused to be created.

In a speech in the early part of 1916 in his own county, H. J. Kershaw, one of the League directors, connected John Y. Gerow, the president of the Dairymen's League, with what he said in the following statement in his speech as reported at the time in a local paper:

"When you have a rotten finger on your hand, it is better to cut that finger off than to let it remain and poison your whole system. When you have an official in your League, no matter who he is, no matter what his position is, if he appears false to his trust and incompetent to hold his office, he should be cut off."

In the same speech he said further:

"I wish to say one thing more and that is though the farmers of Chenango County have raised money through the month of June, many hundreds of dollars, not one cent of it has been sent away. I have told the local secretaries of the leagues to retain that money in their banks until we are sure that the other counties are doing, as we are doing, and that we are sure that it would be spent for the purpose for which it was given and every local officer of the local leagues in Chenango County who is present here today will substantiate what I have told, and all of this money is today held in this county."

And again after advising the election of a new president, he said:

"And then let every present director of the Dairymen's League hand in their resignations. Let this man select in their places men of integrity and influence throughout this milk producing state of ours. A directorate that would have the utmost confidence of this state which the present directorate has not."

In a letter about the same time, Mr. Kershaw wrote that

Chenango County farmers were "openly distrustful of the leaders of the League."

In another public statement he said that the officers of the League were "discredited and distrusted" and that a man like A. A. Hartshorn of his county should be elected president.

The utter helplessness of the Dairymen's League during this period was further officially revealed by Director Kershaw in a statement made in a speech in Chautauqua County at a later date as follows:

On Their Knees

"In 1914 in August of that year, the very month of the beginning of this great world's war, a committee of the directors of the League went to New York, not to demand our rights because they did not have the strength at that time to demand them, but simply went down there practically on their knees as supplicants asking these great milk dealers to give farmers of the State of New York more for their milk so that they might maintain their dairies and continue in the industry which means so much to our state and its prosperity."

"First we went to the Sheffield Farms. These people told us that they would like to give farmers more for their milk, but that they alone could not do so and advised them to go over and see their good friends, the Bordens, but the Bordens would not see this committee, stating that they did not dare to because they were fearful that they would infringe upon the Anti-Trust Law of the state and make themselves liable to imprisonment after they had done so."

Dairymen's Confidence.

The following resolution voluntarily adopted by a meeting of Chenango County dairy farmers (Mr. Kershaw's home county), in the early part of the year 1916 and mailed to me by the secretary of the meeting is typical of many group and personal expressions received at the time from other milk areas of the State:

Chenango County Farm Sentiment

"At a meeting of farmers held in Bainbridge, February 17, the following resolution was unanimously passed:

'Whereas, the market condition for the products of the dairy farms are so unsatisfactory, the prices being arbitrarily made by the large milk companies, that there is little, if any, profit in the dairy business, therefore be it

'Resolved, that it be the conviction of this meeting that the present New York State Department of Foods and Markets is the only agency from which we can expect remedial action, and that the Governor of the State, and the Senator and Assemblyman from this district be urged to use their influence for the continuation and extension of the activities of such Department of Foods and Markets.

'May we ask that any proposal facilitating the marketing of dairy products by the elimination of middlemen's profits and dividends on watered stocks of milk companies, receive your careful and favorable consideration?'

H. H. LYON, *Chairman*."

John Y. Gerow.

During the month of July in that year, 1916, the League directors met in Orange County and requested the president, Mr. John Y. Gerow, to resign, which he did. The directors then elected Mr. A. A. Hartshorn to the place. Mr. Hartshorn, however, refused to accept. When I met him later, he told me that he refused because "they" never had done anything and he believed they never would.

The request for Mr. Gerow's resignation was an alibi for the Executive Committee and an attempt to quiet the criticism of Kershaw and others for failure to do anything, as farmers had said, but "collect money and spend it." When the Department got under way the following October, John Y. Gerow turned in and did his full share of the work.

The Department Had Sole Authority.

Originally I did not intend to have formal contracts of any kind. My program was that after we had succeeded in our immediate purpose of selling the milk at the price we demanded, we could proceed at our leisure to develop a real co-operative system that New York dairymen could direct and collectively protect themselves. But the sale of the milk was the immediate objective. My program was to proceed under the best co-operative ideals. That meant voluntary consent, mutual interest, and the collective good of all. We had no time to canvass for individual contracts, and no time or purpose to try to enforce them if that were desirable. Besides, I felt we were safer without written contracts.

Farmers were trusting me with their milk, and I trusted them.

I came to realize, however, that the instincts and perhaps the judgment of dairymen were better than mine when they advised me to begin with a new organization. There was an element of human nature involved that I had overlooked—pride and ambition. These are worthy human functions provided pride is not allowed to develop into vanity and when ambition stops short of avarice or envy or hate. The small group of League directors, who had become suddenly vocal, seemed entirely unconscious of any selfish interest, but it was clear enough that their activities during June, July and August of 1916 indicated the purpose of one or more leaders to capitalize the sentiment of dairymen that had been developed by the Department during a continuous campaign for twenty months.

They had attempted to affiliate the farm plants and had failed. They had attempted to draw the United States Department of Agriculture into the picture and had failed. They had quarreled among themselves in public. They had approached the dealers "on their knees," as Kershaw dramatically expressed it, and did not get past the office boy in Borden's offices. They had quit when the canvassers, after collecting $50,000, could not collect enough to pay their expenses. They spent $50,000 collecting it. Their venture had been dead for two years but when the Department had revived hope and confidence in organization, they sent out calls for an assessment and collected over $3000, which they spent without any constructive work.

As shown by Mr. Cooper's letter to me, they had no program except the dream of a strike for which they had made no preparation. The stock levy as shown later by the Wicks Committee revealed a loss of confidence and trust on the part of farmers. Whether the new activities were unconscious or intentional, it was all diverting and if continued might defeat the Department's program. I decided, therefore, to clarify the situation in writing so that there could be no divided authority in our program.

I had an early opportunity to do so in a normal way. The directors called another meeting in Albany to elect a president due to the refusal of Mr. Hartshorn to serve. I received a request by wire to attend the meeting. I was not a director or a member. Jacob S. Brill, of Dutchess County, was elected. I submitted a resolution in writing which authorized the Department of Foods and Markets as agent to negotiate the sale of all milk which the Dairymen's League was authorized to sell for its members. This not only avoided any conflict of authority but also served to show that the Department was acting within its charter privileges. It was in harmony with all understandings up to that time and approved, in Mr. Brill's presence, unanimously.

The Dairymen's League was at that time merely a symbol. All field work had stopped two years before. No one knew how many farmers it could rightly claim as members. Of about 12,000 stockholders it had registered in ten years, some were family duplicates, others had died or gone out of dairying because of age, infirmity, or sale of the farm. Many believed it was dead. The new assessment was made at the rate of 25 cents per cow. The amount collected would indicate that no more than owners of 12,000 cows responded. Many paid because it was linked by solicitors to the new hope dairymen began to feel in the activities of the Department. Any dairyman was entitled to the service of the Department without buying any stock or signing any contract.

CHAPTER XII

THE MILK FIGHT OF 1916

A Pep Meeting.

On August 31, 1916, I advised milk dealers that the Department had been designated as the agent of farmers for the sale of their milk to be delivered on and after October 1, 1916; that for ten days preference would be given old customers to contract for their regular supplies in the districts in which they had plants in operation, but no reservations would be made after September 10th. I invited them to call on me, or if more convenient to them, I would go anywhere to sell milk. None of the big dealers replied or called. Many smaller dealers responded and contracted for their milk supply.

To focus general attention on the program, to explain publicly the plans and purposes to both dairymen and the public, and to consolidate the farm sentiment, I proposed a dairy meeting in the city of Utica. The League officials readily approved the suggestion and gave it unqualified support. We had a meeting of about one thousand dairymen there on September 6. In my address I made no appeal to their emotions. I talked to them in plain simple language that farmers understand.

I enumerated consumers, producers and dealers as the three direct interests in the production and distribution of milk. Consumers first, because milk is an essential food for infancy and childhood and because they make the market. Producers came second because there could be no milk without them. Dealers third because they served a useful purpose. But consumers could have milk without them.

It had been asserted that there existed no milk trust. I recited the conditions in the city trade and the way prices were fixed for the consumers, and said:

"There may be no milk trust, but if a milk trust were in existence, these are the things it would do. This is the situation it would create."

I recited testimony from Attorney General O'Malley's hearings and report. I submitted that, "There may be no milk monopoly, but when tradesmen do these things and bring about these conditions of affairs, for the want of a better name I call it a milk monopoly."

I justified an increase in the price of milk because it had been so fully proved that farmers received less than the cost of production. No one rose to deny the producers' loss. I also justified the rise in the price of milk by comparing its food value with that of other foods in daily use, by showing that the prices then paid would lower the quality and supply and so work a hardship on consumers, and I further showed that the State, the people as a whole, and general business would suffer by a destruction of our dairy industry.

I then described the plan that had been devised and largely approved to restore to farmers their rights to set the price on their own milk. This was substantially the co-operative plan already outlined in a previous chapter. I admitted that several attempts by farmers to determine the price of milk had failed, but other organizations had undertaken to fix a price, and the farmer-members were left to sell individually to well organized milk dealers. This time the State Department of Foods and Markets had been designated as sole agent. That put all the milk in one pair of hands to sell, and the price would be the same to all. I urged that "The delegation of this service to a State department is of vastly greater importance and significance than the employment of an individual or a corporate business agency, for the reason that the Department has the backing of the State, the Attorney General and the police powers of the State to enforce its rulings and to protect the patrons, be they small or large, who buy the milk for distribution."

I made it plain that we had no adequate legal form of association and no time to organize one to include all producers. The only authority came from their voluntary words and actions, but the purpose was to develop an organization on a legal framework immediately following the success of our immediate objective, and that it would be so constructed that dairymen would collectively and exclusively control and direct every function of it themselves, that no money could be levied or collected until they had authorized it by ballot vote, or no money paid out or debt incurred until they had authorized it in the same way. I assured them that there would be no secrets, and no mental reservations. Full records would be kept, detailed monthly accounts would be furnished them. I said finally that:

"We have often been told that farmers cannot be induced to hang together. This fiction has been repeated so often that some people believe it. I trace its source to those whose interest it is that farmers should hang separately. Dealers, middlemen, defeated politicians, and disappointed promoters always tell us that farmers will not stick. When the intelligence and good sense of a farming community refuses to be buncoed by smooth and plausible schemes, it seems to be a balm to the feelings of defeated and disappointed promoters to accuse their intended victims of stupidity and want of enterprise. Farmers of the State of New York have had a good many practical demonstrations of the organization of farmers for the benefit of fakirs, and I for one, am glad that the organization of farmers for this purpose is not as inviting an occupation in this State at the present time, as it once was.

"I am one of those who believes that farmers as a class have intelligence enough, initiative enough and loyalty enough to organize themselves into business associations for the protection of their own interests, when the proposed organization is initiated by themselves in a legitimate and practical organization for their mutual benefit. I have seen evidences and instances of this in the past and I believe that the movement in which we are now engaged will demonstrate that farmers have not only intelligence and enterprise to organize a successful movement, but that they have the power and endurance and the willingness to make sacrifices and to pursue a worthy cause to a successful end. If I am mistaken, if the bulk of the farmers refuse to stand together for the organization of their industry, our present movement will fail. If my estimate of farm character and farm spirit is correct, if the old spirit of Bunker Hill and Valley Forge, for principle and sacrifice, are yet in the minds and the hearts of our producing classes, this movement will be successful and we will develop an industry that will be a pride to posterity and a glory to the State."

Dairymen United.

From the very first I detected in the spirit and emotion of the farmers of that audience a determination to see the program through. Simple as the language was, they figuratively made the shingles on the roof rattle with their approval at every period. I had spoken what they had felt and believed themselves and it thrilled them to hear it expressed. Then and there they determined to canvass the dairy areas and hold local meetings to bring producers into the movement. They asked me for the privilege of printing and reading my statement at such meetings instead of trusting to speakers who may not have the information, provided I did not want to reserve it for personal use. I gladly agreed, and printed 100,000 copies which were widely distributed over the State. The motion to use it was unanimously adopted. In recognition of the co-operation of the League officials, I printed an inscription including the Dairymen's League and R. D. Cooper as chairman on the cover page.

I have never known farmers or any class of people who so promptly and generally rallied to a cause of common interest as dairy farmers did from that day and in the five weeks that followed. They travelled at their own expense (we had no money) far and near. If any hesitated because of the cows and the farm work, their wives and children literally drove them out and did the farm work themselves. By this spontaneous enthusiasm we had created practically overnight a voluntary State-wide united dairymen's organization.

Negotiations Failed.

I still had twenty-four days before the contracts expired on September 30th. I contacted the larger dealers but Borden's was the acknowledged leader and many of the others referred me to them. I could not reach Albert Milbank, the president, because he was in Europe. Evidently he had not taken me very seriously. But I discussed the matter for several days with the representatives of Borden's Farm Products, Inc., a subsidiary which handled the New York City business. If they were sparring for information it did not matter. If I had a mirror of my mind and purpose I would have been pleased to have them look at it. But from their manner and speech I felt that there was a good prospect of making the sale.

Finally the spokesman said we had covered the situation pretty fully and suggested that "we all sleep on it over night" and he would advise me next day. The second day following I received a long letter saying they had made a survey of the dairymen, that they found their producers were in good financial state, and satisfied with prices they had previously received. The letter stated that they would send out their schedule of prices as usual to begin October 1.

Dealers Bid Up Price.

Borden's sent out a statement to be posted at their plants offering $1.90 a hundred pounds in the 100 mile zone, ten cents less in the second zone and still ten cents less to producers whose score fell below the barn score fixed by Borden's and inspected and served by their own men. Borden's price for October, 1915 was $1.70 in the first zone. This was a twenty-cent gain already, but they were still twenty-five cents short. Sheffield Farms offered $1.95. Farmers did not respond.

Loton Horton, president of Sheffield Farms, went up to Orange County where we both had many friends. When he spoke of milk contracts, farmers suggested that he "see Dillon." This made him angry. He replied that he "would go out in the adjacent lot, dig a ditch six feet deep, jump in it and draw the dirt in to cover himself rather than go to Dillon for milk."

I always thought of Loton Horton as a diamond in the rough. He was an orphan boy when he left Orange County to drive a milk wagon in the big city. He worked his way up to the head of the second largest distributor firm in New York City. It was no fault of his that he had little education, but when he talked, every one in his hearing knew what he wanted to say. As a milk dealer he acted with other dealers to fix a low price to be paid producers for milk, but I always felt that he had a lingering sentiment for the farm. I was in a fight against him, but personally I admired the good qualities in him. He didn't dig any hole to bury himself in but he did come to me later and bought the milk.

I advised dairymen who had orders for the milk at the price we demanded, $2.15 for October, to continue shipments, and that others having no contract should not ship until we found a market for it.

A Friendly Press.

On October 2, 1916 a group of reporters came to me and asked what had happened to my strike. Dealers had told them they had their full supply that morning. I told them I knew nothing of a strike. Farmers had authorized me to sell their milk for shipment October 1. I had made contracts with dealers who were willing to pay the price. These dealers would get their milk regularly. I was eager to sell it. Dealers set the price of milk for the housewife. If she refused to pay the price, she got no milk. No one called that a strike, and I was selling milk on the same business principle. The reporters saw the point. They came to see me daily, sometimes two times a day for the two weeks following. They said at the end that I had never misled them in a single instance. Their stories on milk were always favorable to the Department and helpful to me.

Who Paid the Gangsters?

I had invited the Executive Committee of the Dairymen's League to come to my office during the disturbance. They came on Tuesday, October 3, 1916 and remained until the sale was consummated, but spent part of the time on matters up-State. There were at the time a goodly number of farm-owned plants in the State. A few of them were under contracts to deliver their milk, but most of their milk was available. These farm-owned plants could handle much more milk than they normally received from their members. They took in all they could handle from producers who formerly sold to the dealers. They were a big factor. The smaller dealers who signed contracts with us got all of it.

Some dealers were friendly. Years before, when N. A. Van Son had the milk of two plants in Delaware County to sell, he, like Mr. Thompson of Holland Patent and others, came to me to help them get a buyer. Van Son had started a business in Brooklyn and another in Manhattan. He, in turn, now helped me. His services were invaluable. The dealers, who bought our milk, supplied their old consumers and broke in on the consumers who had little or no milk. This began to worry the milkless dealers. Toward the end of the first week thirteen dealers came in and each signed a contract. These had country plants of their own. They took the capacity of their plants and proceeded like the farm-owned plants to edge in on the routes of those dealers who still held out. We had then sold about one-half the normal milk supply of New York City.

During the first days of the trouble I received complaints that a gang of "toughs" followed the delivery wagons containing our milk in the early morning and dumped the milk left at the door before the store was opened. They damaged the cans and threw the lids in places out of sight. I went to see the District Attorney of New York County. He was apparently indignant and said he would "drive knives into them and twist the blades in their flesh." He did nothing.

I then went to Attorney General W. W. Woodbury. He stopped it quickly. But we next found the gangsters lifted the lids of the cans and dropped in tablets that soured the milk. The Attorney General made some arrests and stopped that. Another complaint came from drug stores using our milk. City inspectors visited the stores four times a day, annoying the proprietors and arousing the suspicions of customers. I do not know how the State Attorney General handled these complaints but the practice was discontinued. I never discovered who paid the gangsters for their work. It could not have been dealers because their great alleged worry was that babies might not have milk!

Investigation Denied.

After the foregoing experience the big dealers determined to turn the services of the State Attorney General's office against me. They made a formal complaint alleging that I had exceeded my authority and violated the law, and they demanded action to

THE MILK FIGHT OF 1916

remove me from office. At the same time they brought a personal suit against me in the Supreme Court for damages of $150,000, which they alleged they had sustained in their business through my activities.

Sometime in the latter part of September I had written Attorney General W. W. Woodbury requesting an opinion on a proposition I was considering. To lay a ground for his opinion I gave him a statement of the milk situation in the State. When he received the dealers' complaint against me, he interpreted my letter as a complaint against the dealers and stated he would investigate both cases together. A Master was appointed by the court to hear the evidence.

When the hearing was opened in New York, the dealers strenuously insisted that the investigation of me should come first. Mr. Woodbury calmly said he would take the dealers' case first. He wanted to examine the dealers' books to determine the cost of distribution. The dealers' attorney objected and the Master sustained the objection. The hearing closed. Then my accusers joyfully demanded that the investigation of the Commissioner of the Department of Foods and Markets proceed forthwith. Mr. Woodbury replied that he was not going to examine the Commissioner. "You have," he said to the dealers, "instituted a suit against Mr. Dillon for a large sum of money and I am not going to use the Attorney General's office to help you dig up evidence against any citizen in a civil suit." That closed the incident. The civil suit was finally dropped.

By this time the tension was high. My office telephone was busy every minute of the day. At home I was called out of bed late at night and early in the morning to answer telephone calls from both nearby and distant points. Orange County farmers called a rally meeting. I was asked to come and address the meeting. Friends pleaded with me to come. I had interests there, my friends were there. I was more than anxious to go, but I just did not dare take the chance of being absent even for one day. To me it was a real sacrifice, but I remained at the post. Farmers frequently dropped in to see how things were going. They were naturally anxious, but they stood pat. They worked among themselves to make the best use possible of their milk, and to persuade other producers who had hesitated to join them.

By the beginning of the second week the big dealers became desperate. Most of their plants were stone dry. They were getting only about fifteen per cent of their regular supply. They were paying $2.40 a cwt. for what milk they could get from out-of-State. Sheffield Farms increased their offer to farmers, Borden's increased the price twice in one day. Neither of them got any new milk. Borden's sent a wagon out to collect milk from Earlville; it came back with empty cans.

A Sole Disturbance.

I cautioned at the start that any report of violence would hurt our cause, that our success depended solely on the voluntary support of dairymen, and that if, by our conduct, we merited confidence it would come to us, and few, if any farmers would fail to join us.

The only disturbance that I recall was an argument on a road in Dutchess County. During the early days a small group of dairymen gathered to persuade a neighbor not to ship milk. The dispute developed some heat, resulting in the dumping of a can or more of milk. A complaint was made to John Mack, who was then District Attorney of Dutchess County. Mr. Mack called on me to find out what it was all about. I assured him that the complaint must be the result of a local incident. I was selling milk in a business way and as a haberdasher put the hat back on the shelf if the customer refused to pay the price, I held the milk on the farm until the dealer agreed to pay our price. Mr. Mack intimated that he believed that was considered a good business practice. He has received high honors since and we have remained friends.

As a matter of fact, the dealers' attitude was more in the nature of a strike against farmers. The dumping of our milk, the destruction of our milk and cans and the adulteration of our milk were acts of violence. The annoyance of the customers who bought our milk was deliberate, and the boycott later of the small dealers and storekeepers who bought our milk was unlawful and revengeful. No farmer or farmer's representative had any wish to harm any milk dealer or his property.

Stunts and Conspiracies.

Then they started a series of new stunts. The consumers, the press and the public were clearly against them. They attempted to turn the tide of sentiment against us. They made the charge that we were depriving babies and hospitals of milk. I replied that if they would show that any child or individual or hospital needed milk in excess of our supply, I would furnish free milk and the dealers could pasteurize it. As a matter of fact, the children and hospitals were fully supplied.

The next stunt was sinister, but the hysterics were laughable. Borden's threatened to build stables in Jamaica, L. I., and supply their own milk. I offered them the use of a good producing farm of 300 acres in Orange County if they would run it for a year, pay taxes and interest on the investment and produce milk at our price without loss, and I would let them keep the profit, if any. The offer was not accepted. Borden's had run one Orange County farm.

A Frame-Up.

One night at 10:30 P.M. my telephone rang. Mr. Cooper was calling. He, Mr. Thompson and Mr. Manning had been talking with a lawyer named Breed, who had a proposition to settle the milk dispute. He was urgent, "Would I come down to the Manhattan Hotel to meet him?" My wife protested, but my attorney, who had dropped in a few minutes before, offered to accompany me, and she relented. Mr. Breed told me his story. I learned in the course of it that Mr. Breed had met the Mayor quite by accident that morning. Later he met purely by accident Dr. Emerson of the City Board of Health. He also on the same day met one of the Borden's executives by accident. In the interest of the public they were all anxious to settle the milk dispute. In fact the Mayor, he said, was quite insistent on it. It had been proposed to refer the matter to a committee consisting of eleven persons, one to be appointed by the State Grange, two by the

Commissioner of Agriculture and eight to represent the Merchants' Association. To authorize this, a meeting in the Mayor's office was proposed for the next day at noon.

I was not impressed by the number of significant accidents Mr. Breed had in one day; but that aside, I considered the Merchants' Association disqualified to sit on any disinterested committee in which farmers' interests were concerned because it had already taken sides against farmers' interests and in favor of middlemen without any attempt to look into the merits of the farm case. I related this to Mr. Breed and considerable discussion followed in which I made it politely plain that I could not consent to arrangement for discussion of a question that was already decided against me. It was clearly a frame-up. As I rose to go, I stopped to talk to Mr. Thompson and heard Mr. Cooper say: "Is it your idea, Mr. Breed, that you would represent us at the Mayor's conference tomorrow?" Mr. Breed expressed consent. I assumed Mr. Cooper had fallen into his trap. It was just what Mr. Breed had seemed to be fishing for since I entered the room. Cooper had, I thought, unconsciously and stupidly asked the question that Mr. Breed had covertly suggested. I promptly left the room. I felt speechless. Neither of us spoke until we reached 70th Street and Broadway. Then my companion said, "Don't you want to stop at the hotel and squelch that Cooper-Breed appointment over the phone?" I said, "I am going to bed." The only other words spoken in that early morning ride were two "Good nights" as I left the cab at my door.

The next morning I explained the situation to Mr. Thompson who was not always quick to see the significance of an act, but he was an honest and sincere man and when once convinced he could be depended on to stand by his convictions. I convinced him that Breed meant mischief and that it was his duty to see that Cooper called up Breed and cancelled the arrangement of the night before. I also put it up to Mr. Cooper and insisted that he break with Breed. It took him forty-five minutes on the telephone to get out of it, but he did.

In the meantime I had received the Mayor's invitation to his conference. While waiting in the outer room Dr. Emerson, Com-

THE MILK FIGHT OF 1916

missioner of the City Board of Health, came to me saying that because of the dealers' opposition to me, it would be best for me to consent to say nothing and allow the conference to fix the terms of settlement. I replied that such a procedure might be consistent with his position and satisfactory to the milk dealers, but it would not be in keeping with my duties to the Department and would be a violation of the trust I accepted from dairy farmers.

The conference with the Mayor was short because I made it plain at the start that I would not be interested in a loaded committee of eight to one against me at the start. That ended that stunt.

Three More Tricks Failed.

After this conference with the Mayor, there came an offer to pay our price if I would consent to allow dealers to deduct their loss on account of what they called "surplus." They could hardly have expected me to fall for that. The offer was intended to make me appear unreasonable and bull-headed. The fact that they expected me to overlook was that in 1915 when I suggested one cent a quart increase I took Borden's prices as the base because Borden's prices had been generally accepted by the trade for some years. This was a monthly flat price to be paid farmers each month for a year. The high price in the schedule was in the months when there was little if any "surplus," and the price was less for other months in proportion to the estimated "surplus" for each month. To accept that offer, therefore, would be to make double deductions for "surplus." When I made this plain that trick faded out.

Then for the same purpose of trying to make me seem stubborn and indifferent to the public inconvenience, an offer was made to pay the flat price as demanded up to January 1 and leave the price for the next three months to be determined late in December. That was a clever thing for them to do. They were willing to pay the price I asked, and if I refused this time I certainly would be condemned as an obstructionist without reason.

That evening Attorney General Woodbury had all the afternoon city papers laid out on his table with the headlines showing when I called at his request. He wanted me seriously to consider the latest offer. He reminded me that the press of the city had been favorable to me for nearly two weeks. There was no criticism of me yet but to him there seemed to be a favorable attitude to the offer in the headings. He related several things that might happen to my cause and to me personally if the papers turned against me.

He was my strongest support through the whole ordeal. His sympathy with the farm was sincere. I was conscious of his interest and his friendship. If it concerned myself alone, I would not hesitate to demonstrate my appreciation of his good interest and services, but in this an industry and a people were concerned. I expressed something of this to him and at his request consented to consider his advice seriously over night.

I was not concerned about any effect the decision could have on me personally. I saw clearly the dealers' purpose. They would get their requirements for the three low production months of the year. They would then demand a lower price for the next three months. We would have a choice of accepting their price or going through the same thing all over again. It would be a compromise, not a victory. The timid would be glad to have it over, but it would be a great disappointment to our staunchest friends. Above all, we would have compromised our right to fix the price and terms for our own product—the great principle we hoped to defend.

The next morning I told Mr. Woodbury that if I was sure rejection of the offer would cause my brains to be knocked out of my head with a club before noon, I would still reject it. The big maneuver had failed, but they had a last trick in reserve.

The night after I had refused to accept the last offer, Mr. Brill, who as far as I knew had taken no part in our milk-selling program, sent telegrams to dairymen up-State saying that the dealers' offer was accepted and directing them to ship their milk.

This created a serious situation. He could reach only a small number of shippers, but his action was reported in the morning papers and confusion was sure to follow. The dealers would, of

THE MILK FIGHT OF 1916

course, make the most of it. Mere arguments would do no good; I had to demonstrate that Mr. Brill had no authority to act for dairymen in or out of the League and that I had. The League Committee agreed to my program.

At my suggestion Vice-President F. H. Thompson sent out telegrams to producers at once to ignore Brill's wire and hold the milk. He also called League directors to a special meeting on Wednesday night in New York. My lawyer drew up a resolution directing Mr. Brill to acknowledge that he did not have authority to sell the milk, and to withdraw the telegrams. In the event that he refused to comply, another resolution declared the position of president vacant and still another to delegate the president's powers to the Chairman of the executive committee. The next morning this procedure was reported in the city papers. None of the milk reached the city.

The Agreement.

On Friday forenoon my attorney and the dealers' attorney arranged a meeting with the dealers for 8 P.M. at the Sherman Square Hotel. All important dealers were represented except Borden's, who, however, sent word that they would accept any settlement made at the conference. It lasted until 4 o'clock the next morning. The biggest obstacle was the advice of the dealers' attorney not to sign a collective contract because of the anti-trust laws. I agreed to rely on their word provided we made identical memorandums to avoid misunderstandings.

Another obstacle was the dealers' alleged fear that supply and market conditions would not justify the extra price for January, February, and March. I had suggested the increase more than a year ago. In the meantime the European War had rapidly increased the demand for dairy products. The O'Malley investigation six years before and the Wicks hearings during that same year—1916—had shown that milk was being produced at a loss.

George Warren of Cornell had shown that farmers were working at low hired men's wages without any pay for family help and no profit. Prices of labor and commodities had gone and were going up. It was no secret, the dealers knew it as well as I did.

They were there to settle but they were trying to save their faces. I was willing that they should, provided it cost us nothing. I insisted again that all we wanted was cost of production and a reasonable profit, but the six months' price must stand. If it could be shown that our price for January, February and March exceeded our objective we would readily consent to a modification of the price for these three months to correspond with the cost of production and a reasonable profit. This was finally agreed to. We provided for a committee of two persons chosen by each party and, if needed, a fifth person chosen by the first four to determine the facts as to cost of production. It turned out that the committee of four agreed to an increase of an extra five cents a cwt. for the three months.

We were then in agreement. Messrs. Cooper, Thompson and Manning were with me. I conferred with them and they approved it all. Everybody felt happy. Dealers who had seemed a bit bitter during the scrap came to me and commended me for making a clean fight. They all wanted the milk as soon as possible and we lost no time in getting it to them.

All of the commercial milk of the State was sold at a flat price for six months at an increase in the price of forty-five cents a cwt. with a 5 cents increase for three months. It was renewed for six months. It was the first and last time in a period of forty-six years that New York dairy farmers collectively and exclusively sold their milk for a definite period at a price fixed by themselves.

Saved by Prompt Action.

If I had not acted promptly, the Brill episode might have resulted in our defeat. I had met him only twice; both meetings were brief, and the price and terms for the sale of milk were not discussed by us on either occasion. But it seemed that he had disputes with the other members of the Executive Committee at the start. As a newcomer, Mr. Brill, it seemed to me, took his position too seriously. True, we had adopted the name of the corporation for the machinery, but while Mr. Brill magnified its possibilities the other members well knew that the existing stock members were trifling in numbers compared with those who had

THE MILK FIGHT OF 1916

come in voluntarily and informally on the appeal of the State Department. They also knew, as previously shown, that they did not have the confidence of the existing stockholders. It was Mr. Brill's chief misfortune, however, to have had two good and sincere friends who like himself had accepted the advice of a farm paper editor and his employer, and these, blinded by self-interest and envy, led him into the councils of the milk dealers. It was this combination that induced Mr. Brill to send the telegrams, which, if successful, would have defeated the farmers' program. Later, his two sincere friends realized their mistake and not only admitted their error but did much to atone for it in a modest practical way.

I confess that at the time I was puzzled by Mr. Brill's opposition. I had then worked forty years for farm co-operation. During the two years just passed I had talked milk all over the State of New York and in every section of the metropolitan city except Staten Island. I talked it at club lunches and dinners, in schools, and forums, in private homes and in public schools, and in churches from humble Hester Street to fashionable Fifth Avenue's notable pulpits.

I was devoting the editorial pages of my paper to this work incessantly and was using the prestige and power of a State Department to create an organization of which Mr. Brill was president, and trying to put it on its feet that it might atone for the hardships of departed kindred and living friends. This was not for hire: I was working for nothing and boarding myself. My State salary was diverted to the expenses of the Department. Certainly there could be no offense in all that for a disinterested friend of dairy farmers.

After it was all over I became better acquainted with Mr. Brill. I came to understand the cause of his dispute with the Executive Committee and to realize that he was unfortunate in the leadership he accepted. He had been ill-treated by his own committee, and misinformed and misled by the leader of a small group who fanned his resentment for their own purposes. When I came to know Mr. Brill, I found him an agreeable gentleman, and we remained good friends during the rest of his lifetime. He passed away in 1937.

The Farm Bureau Boys.

Most of the Farm Bureau managers at first hesitated to take part in the milk organization program. They were organized originally to help overcome the "high cost of living." To the proponents of it, that meant reduced prices for farm products. Commercial concerns made contributions to the bureaus and local distributors became members and made contributions. Among these in New York State was the Borden Company. Then distributors of farm products and of supplies sold to farmers, strenuously objected to the competition of farm co-operatives either in buying or selling. They insisted that the Farm Bureau, which they helped support, should not assist their competitors.

To meet this difficulty, Prof. M. C. Burritt, Director of Farm Bureaus for New York, instructed the managers "to avoid interfering more than is necessary with established business relations" and to limit their services to farm associations to furnishing information and "never have any business connections with them after they were organized."

In Dutchess, however, County Agent F. H. Lacy jumped into the work from the start and kept going to the end. In Cortland, County Agent E. H. Forristall sent word to his superior that he had gone into the work, that he couldn't stay in the county if he hadn't, and besides it was just what he wanted to do. In Orange County, Tom Milliman had helped locate a chocolate milk concern the year before. Borden's objected on the ground that they contributed $250 annually and the agent should not interfere in their territory. But when the fight got hot, Tom felt the embarrassment of his position. He went to Cornell and was finally told that he could use his judgment about it. Soon the Farm Bureaus in most of the dairy counties were in the work in shirt sleeves and did helpful work, and it popularized the "boys" in their counties.

Reward in Thrills and Friendships.

Aside from the thrill I felt in closing the sales, I prized most

THE MILK FIGHT OF 1916

dearly the many friendships I had made during the two years' preparation and execution of the program. Some of the most delightful friendships developed by correspondence with men and women whom I had not met. Many of them I never did meet. No small number of them have gone to their rewards. But more than a hundred thousand milk producers in the New York milk shed were either active in the struggle or sympathetically watching it. I meet many of them still and hear from them often.

During the fight and especially at the end, men and women sent me in prose and verse congratulations and assurances of appreciation. They were so generous in their appreciation of my part that they forgot entirely the greater part performed by themselves. I had simply happened to be in a position where I could give them the leadership necessary to their success. Now, as they refer even casually to the old emotions and trials, except for family sentiments, they revive in me the most cherished memories of my life.

I am not entirely friendless among the milk dealers who were our opponents in that memorable struggle. I knew they would keep their words in the agreement. They did. They know the interests that I represent, and respect them. Personally, there were many likable men among them, but they inherited a system with unfortunate traditions, and I suspect it is as difficult for them to change as it is for farmers to correct it. Their interests and those of farmers are antagonistic only in the matter of price. If politics could have been kept out of it, I believe I could have bridged that difficulty.

At no time before the real struggle began until the end of it was there any difference in word or purpose, as far as I knew, between myself and any officer or member of the League. None of them ever interfered with anything I did, or questioned my authority to do it. They were guests in my office at my invitation during the fight and they scrupulously respected my position as head of a State department. They respected my sense of responsibility and gave me no cause to question their sincerity and loyalty. We were all eager to popularize the organization because the name with a new co-operative charter was to be the symbol of farm unity for the future. For that reason, as well as to show that I was keeping within my legal limits, we all kept the Dairymen's League in the foreground.

What I have said of the limited number of the old officers and members of the League was equally true of the great mass of dairymen who came into the association through the Department. No word of dissent from them ever reached me. On the contrary, they gave me many touching evidences of their confidence. Their support was unanimous and inspiring. If the fight had cost me my life savings I would have deemed the victory and its possibilities worth the cost.

A Significant Proposal.

We estimated that one cent a hundredweight would furnish sufficient funds for the expenses of the organization. The collection of these funds was a problem. It was suggested to ask the dealers to reserve it on the producers' monthly accounts, and turn it over to the organization. I did not like that method; first because I liked to preserve the voluntary policy which had worked so successfully, and second because I felt a monthly check from the dealers would have a reciprocal response from the management, and in time cause the officials to forget that it actually came from the producers. However, it was the expressed understanding that the association would be promptly organized formally, and I consented to the plan of collecting the funds in the meantime through the dealers. Future events tended to show that my objection was sound, and convinced me that yielding even for a time was my major mistake up to that time.

After the funds commenced to come in, it appeared that the potential revenue would amount to about half a million dollars a year. In the meantime, the Executive Committee having taken offices of their own, came to me and Mr. Cooper said we would need to change our plan. I asked in what respect, and he said that it gave too much authority to members and not enough to the management. "For instance," he said, "farmers cannot be trusted to make rules for the delivery of milk to plants; they would bring it any time before noon. Then there is another thing, farmers have never been in the habit of paying salaries. They would expect men in the association to work for hired men's wages, and if 'we' [the Executive Committee] were to do the work, we would have to be paid for it."

In answer I said, "For forty years other men have run the farmers' milk business. Farmers didn't like the system. Now they want to run it themselves. You men were at Utica when I presented the plan. With all the others, you approved it. You and I have stressed it ever since. It was the spirit of it that rallied the farmers of the whole State to it. I could not change it now because it would be a breach of good faith to the dairymen who relied on my promises. Besides," I said, "I believe the principles are fundamental and sound, and I would not change them if I could." This was all said in a deliberate and friendly spirit. He did not dispute what I said, but seemed to be convinced.

CHAPTER XIII

ORGANIZED FARM CO-OPERATION

Co-operative Principles.

At this point in this dairy narrative, I think it worth while to state the needs, purposes and principles of farm co-operation. Before the Civil War, America was an agricultural country. Farmers practiced diversified farming. They lived largely on the products of the farm during the growing season and stored food in the cellars and in barrels to keep the family well-fed in the winter. They sold their products to local tradesmen and local consumers, and in the latter case went home with 100 cents of the consumer's dollar. They bought their supplies at the country store and their implements from the local craftsmen. Much of the trade was barter; the products of the farm were exchanged for the products and services of the shop. There were no middlemen in these transactions.

After the Civil War the country turned to large manufacturing enterprises. The things that had been made by hand in local shops were made by steam power in the factories. Large centers of population were created. Food was shipped from the farms to feed the factory operators, and raw materials were shipped for other purposes. A middleman system sprang up to facilitate the sales. The farmer shipped his produce to the city and accepted whatever the middleman chose to return, if anything. Another line of middlemen shipped the farm machinery and family supplies to the country, and the farmer paid what was demanded. The farmer had nothing to say about the wholesale price of his product nor about the retail price of his farm and home supplies. The middlemen took advantage of their position. The distributors were organized and worked in units. Farmers were not organized. They bought and sold as individuals. The farmer was robbed in the wholesale price of his products. He paid high retail prices for what he had to buy. Co-operation was devised that farmers might protect themselves collectively by means of organization. The formula was:

First: Voluntary association.
Second: Equal voice of every member in the management of the association.
Third: Equitable share of the benefits by each member, according to his contribution to it in work or trade.

It was clearly understood at the start that farm co-operation would be an institution for farmers only. It was to restore the bargaining power they once had enjoyed and had lost. No middlemen were to be allowed to come in or to develop within. It was to be self-help and mutual help. In the late system middlemen did the bargaining to make profits for themselves. It was not satisfactory. Now farmers would again make the bargains themselves, collectively. They may employ help but the information, the judgment and the responsibility were to be their own. To turn the work over to others would be to create another set of middlemen between themselves and the old middlemen, and double their troubles and expenses. There can be no real organized farm cooperation unless farmers reserve full control and do it themselves.

Where farmers of a community organize a local co-operative association to improve their conditions and powers, every farmer can have full information first hand. This enables him to do his part efficiently and intelligently.

Lawyers are schooled in business stock corporations. In these every share of stock has one vote. Usually one man or a small group have shares enough to control the corporation and the business it does. The purpose is profits on the capital invested. The large stockholders are usually in the management and either run the business or direct the executives.

The purpose of a farm co-operative is not to make a profit in the business, but to create a mutual service for its farm members. Therefore, a membership corporation is better adapted to farm co-operatives. The benefits are not paid in dividends but in the service to members and in better returns for their products. It is an organization not of capital but of men, and each member should have one vote in its control.

The strength of a farm organization is in its unity, and to maintain unity all members must share equitably in its benefits. If individuals or groups have advantage over others, unity is destroyed and failure results. It would be folly to authorize a man to vote on a business proposition and deny him full information as to the finances and details of the business. If co-operation is to be successful, the members must have full and detailed information. The rightful interest of every individual in the industry, whether he be in the organization or outside of it, must be protected. Favor some groups over others and you divide your membership. Unity is destroyed. Try to build a fence around a market for your own members, and the excluded or neglected producers will tear it down. They will compete with you and you will be responsible for the chaos. Unless leaders have the spirit of mutual help there can be no co-operation, and there is sure to be a return to competition or monopoly and ultimate failure.

The tendency of all organization is to form minority "rings" or groups for official control. These cliques perpetuate themselves in the management. Selfishness and often corruption follow. Co-operatives are no exception. To avoid this, rotation of the officers is advisable, with one-year terms and no officer to succeed himself for more than one consecutive term. This is sometimes criticised because experience is lost, but if a retiring officer of a co-operative refuses or neglects to give his successor the benefit of what he has learned in office, it was a mistake to have elected him, and a bigger mistake to re-elect him. Rotation of officers develops a wide range of talent and widens the interest and confidence of the membership.

The charter and by-laws of co-operatives should provide for a ballot vote for the election of trustees and executive officers. This is a check on the tendency to form minority groups and helps maintain membership interest and confidence. The ballot vote should also be required for the authorization or collecting of funds, for fixing the amount of expense budgets, and to determine all propositions affecting the responsibility of the whole membership.

That he may vote intelligently and effectively, the farmer must have full information in reference to the business and finances of his co-operative. The charter and by-laws, therefore, should provide for the distribution of detailed information to the members and guarantee the members free access to the books and records as well as a monthly profit and loss accounting. The neglect of these formalities has caused many disappointments and even calamities in attempted farm co-operation.

CHAPTER XIV

UNITED DAIRYMEN HOLD A MEETING

A Made-Up Slate.

The first mass meeting of dairymen since the six months' price was determined was held at Utica, New York, on December 8, 1916. Dairymen individually and in groups came to me to talk about officers to man the organization. John Arfman, of Middletown, who had been an enthusiastic worker during the price struggle, appreciated the need of a capable head. He proposed a successful breeder of Onondaga County for president. Albert Manning of Orange County was the secretary of the League since the beginning. He was not an executive type, but was honest and sincere. I believed he would make a trusty and efficient secretary. L. M. Harden, Sussex, N. J., had been treasurer. He was a good man, but for the sake of economy it seemed to me that for the present Mr. Manning might well be entrusted with the work of both secretary and treasurer.

The embarrassing part of it was that the directors during the Brill episode had delegated the powers of the president to R. D. Cooper. There were such men as Frank Sherman, of Copake, N. Y.; Harry L. Culver, of Amenia, N. Y.; J. D. Beardsley, of New Berlin; and Oscar Bailey, of Brewster, who had connections with the League, and a wide circle of able men who had come in recently through the Department. These included such outstanding men as Fred Boshart, of Lowville; Fred Sessions, of Utica; Hugh Adair, of Delhi; H. E. Cook, of Lowville; J. Leslie Craig, of Canastota; Frank Brill, of Canastota; and Clark Halliday, of Chatham. I do not know how many of these could accept the place, but the list of qualified men could be extended indefinitely. The executive committee appeared at the meeting with a slate made up, R. D. Cooper, President, J. D. Miller, Vice President,

Albert Manning, Secretary and M. W. Davidson, Treasurer. John Arfman was there with a following to propose a prominent producer for president. There was some lively debate during the forenoon. Jacob S. Brill was there with a very small group, but got no recognition whatever. The dairymen were in good spirits. After forty years they had triumphed in a bitter fight which lasted two weeks. It was not the paltry forty-five cents a cwt. extra for milk that they rejoiced over. They had thrown off the yoke of tyranny. They had discovered their own power. They realized a new-born confidence in themselves. They felt the inspiration of hope and trust for the future of their dairy business. They felt full of good humor. They were in no disposition to deliberate on principles or policies. John Arfman caught the spirit and concluded, as he said, that the most important thing was that nothing be said or done to mar the harmony, so the official slate went through.

Co-operative System Ordered.

On the morning of the convention after I arrived by sleeper, Mr. Thompson came to me, wrapped his friendly arm around my shoulders, and said Brill was in town and the committee feared he would make trouble; they sent him to ask me to say nothing about the plan of organization, as it might afford an opportunity for Brill to stir up trouble. I said I did not believe Brill would get very far with that crowd as a trouble-maker, but in any event he would get no provocation from me. I took a seat in the rear of the convention hall, and spent the day greeting friends as they came my way.

When the program was finished and adjournment was expected, a large, square-shouldered sterling type of farmer arose in the middle of the room and said the pleasure of the day to him would not be complete until he heard from Mr. Dillon. Would the chairman please call him to the platform? After the roar of approval subsided, Mr. Cooper said, "Yes, Mr. Dillon helped us."

My first thought was to follow the mood of the day, to say something pleasant and make it short. As I turned to the audience, however, I saw by the faces before me that they had had enough of the emotional talk. It was plain that they expected something more serious and concrete from me. I told them as plainly and briefly as I could that there was no cohesion in the organization; that it was like a chain of sand ready to break at any point, and that the enemy was well organized and had already threatened to overdo us. I recited the fundamentals of the organization that had been approved three months before, and recommended that they appoint a committee then and there to prepare a charter and by-laws to put the proposed plan of organization in legal form.

A motion was promptly made to provide a committee of seven, with Mr. Dillon as chairman, and the other six members to be appointed by the president. An amendment was proposed to refer the matter to the board of directors. After brief debate the amendment got six "ayes" and a roar of "noes." Then the original resolution went through with an enthusiastic vote.

I asked Mr. Cooper if he would confer with me before he appointed the six committee members, as it would be wise to have a strong committee representing all groups of dairymen in the State. The next week he called a directors' meeting in the State of New Jersey and passed a resolution against organizing, thus recording official repudiation of a unanimous act of the membership. Without consulting me, as he promised to do, Mr. Cooper appointed six men who opposed organization. I personally drafted organization papers, but during the whole year following the other members could not find time to consider them. It was easy for a small official group to control a New Jersey stock corporation with proxy privileges.

Rule or Ruin.

That New Jersey resolution usurped the authority and powers of dairymen. It was an impudent official repudiation of a resolution adopted deliberately and unanimously by a convention of the representative dairy farmers of the State. It was the first open, intentional, autocratic challenge of the principle and purposes of farm co-operation in America. It was noticed that the Cooper group was determined to create an official centralized authority over the New York dairy organization, and that they would split the organization, if necessary, to win their purpose. They were determined to rule or ruin.

CHAPTER XV

FARM LEADERSHIP WEAK

The Towner Bill.

During the 1917 legislative session Senator Towner of the Dutchess, Orange and Putnam Senatorial District, introduced the Department's milk bill which Senator Brown had side-tracked the year before. The bill asked for an appropriation of $300,000 or "as much as may be needed," to build a milk plant at New York City to facilitate the sale and distribution of milk and cream direct from farmers as already defined. The purpose was to demonstrate the necessary cost of city distribution of milk under an efficient and economic system.

It was proposed to pay all expense, taxes, interest on the investment, cost of production to the farmer and a reasonable profit to amortize the original cost of the plant. The plant was to be operated in the open. Any person could visit it and inspect it, and the open books would show every item of the business transacted. It would not be big enough to affect the dealers' business. If it failed, as the opposition to it predicted, it would simply discredit my policy and the department. If it succeeded, it would definitely fix the reasonable cost of milk distribution.

When I had proposed this plant two years before it was approved by dairymen generally, individually and in groups. It was a part of the original 1916 plan, and indorsed by old and new members. It was fiercely opposed by the big milk dealers and their political lobbyists. The first canvass of the Senate showed 38 Senators for it. The others said they had not yet read it. There were votes enough to pass it in both houses, but it was held up in committee. Senator Towner, than whom New York dairymen never had a better friend, moved to discharge the committee and bring the bill to a vote in the Senate. Senator Elon Brown,

FARM LEADERSHIP WEAK 113

its original enemy, fearing a defeat, exercised his official prerogative as Senate leader to suspend the rules, force a recess, and call a party caucus. In this caucus Senator Towner forced a promise that the Finance Committee would consider the bill and make a report within a week. It was, however, delayed until all bills went to the Rules Committee and pigeon-holed there to the end of the session. The strategy of a minority defeated it.

City Price Up 2 Cents.

Following the October, 1916, settlement with producers, Borden's increased prices of B Grade milk to consumers from nine cents to ten cents a quart, from eleven to twelve cents for Grade A, and for cream from fifty-six cents to sixty-four cents. During the first week of February, 1917, Borden's again increased the price of Grade B milk to consumers one cent a quart, making the price eleven cents, and cream to seventy-two cents. No increase to producers was made at this time.

The editorial pages of the *New York World* commented as follows:

"The Borden Company raises the price of B Grade milk to 11 cents a quart. Because he sees that 'the present high cost cannot last', Loton Horton seeks to unite several smaller dairy companies with the Sheffield's to economize in distribution.

"Commissioner Dillon thus analyzes the February cost of a 40-quart can of milk: Farmer, $1.74, freight, 34 cents, and pasteurization, 15 cents. The consumer pays $4.40, of which the distributor takes $2.17; all other costs combined being $2.23. Grade A milk costs $2.53 of the consumers' $4.80.

"What has happened is this: The farmer gets practically one cent a quart more, the reckoning being by weight. The distributor takes another cent for himself, charging the consumer two cents more than last year. For 40 quarts of Grade B milk he gets 47 cents more (of the increase) than the farmers with their vast investment, heavy labor, high costs of feed, and cartage. For 40 quarts of Grade A milk he gets 26 cents more (of the increase) than the farmer, the railroad and the pasteurizer together.

"Why should this charge be necessary? If we accept Borden's statements as to losing money more rapidly, though taking half the heightened cost, how can we account for such an amazing fact? Why are they losing more money though exacting a cent a quart more for their services? What other explanation can there be than Mr. Dillon's—that the system of delivery is a costly failure? Mr. Horton's experimental economics should be worth watching."

What the *World* means to say in the last paragraph is that Borden's exacted two cents a quart from the consumer because of a one-cent increase to the farmer over the prices of the previous year.

Consumers Fear Conspiracy.

Following the Borden's February 1917 increase in the price of milk to consumers, the milk subject was fiercely discussed all over again. The League Executive Committee ran some advertisements in the city papers. The city people felt that the advertising indicated a change in the farmers' policy. The advertising seemed to agree with the dealers and to justify the increased price to consumers. The original farm plan had assured them that milk dealers would not be allowed to overcharge consumers. There would be a new negotiation with farmers for April 1st, and consumers feared another rise. They urged the city papers to do something about it.

One morning about the first of March, 1917, the reporters of the city papers started to the District Attorney's office to demand an investigation of the League and the milk trust. Discussing the subject on the way they decided first to come to see me. They told me that the League officials and the milk dealers had been holding secret meetings and neither of them would give the reporters any information as to what it was all about, and that they had come to me because I had always been frank with them. I made excuse for the League officials because of their inexperience, and because the dealers' leaders had misled them, but I assured the reporters that farmers had not changed their policies, and that they were yet opposed to unreasonable increases in the price to consumers. I asked them to let the matter rest for a few days and I would try to get some definite information for them. They did so, and I kept my promise.

The next morning I had a telephone call at my home from John D. Miller of Susquehanna, Pa. He said he was in New York, that

FARM LEADERSHIP WEAK 115

the "League boys" were in trouble and he wanted to see if I could help them. I said, "Bring them to my office in an hour." Mr. Miller had called on me during the October milk fight with a letter of introduction from his son, whom I had befriended some time before. I had freely given him the milk information he wanted. He had gone to the Utica annual meeting in December 1916 with a small delegation, and had become the director from his district.

On this occasion Mr. Miller and the League Executive Committee kept the appointment at my office. R. D. Cooper, the chairman, said to me that they had not been able to renew the October contract with the dealers for the second six months of the year; that, as I knew, the metropolitan dealers had created a strong organization which they called the "Milk Conference Board" and they were very stiff and unyielding. He said the committee had come down to see if I could help them out, and, if they had to "strike" and hold the milk, would I lead them with the Department as I did before?

I said, "Yes, I'll help if a fight is necessary for the interests of farmers. That is what the Department is for and it is what I am giving up my time for. But first I want to give you boys some advice. You are flirting with evil spirits and you are leading straight to perdition. You are eating and drinking with these dealers, and you are no match for them. When you get up from their tables they know everything in your system and you don't know a single thing that is in their heads. Stay away from them until it is time to sell milk, and then sit at opposite sides of the table."

Then I said, "Your present trouble is due to the fact that your friendly visiting with the dealers and your advertising seem to indicate that you approve a two-cent a quart increase in the price to consumers, and the suspicion that another increase may follow. You must show the consumers and the press that such is not the farm policy or farmers' intention."

I then suggested that the Executive Committee go back to its office and draw up two resolutions which I outlined:

1. Approval of the Towner bill, then in the Legislature, providing for

a city milk plant to show that milk could be distributed for from 2 cents to 4 cents less than consumers were then paying;

2. That the farmers' price scarcely covers cost of production and that the farmers' plan contain this provision: "Dealers must not exact unreasonable prices from the consumer and thereby reduce consumption and cut off our outlet for our milk."

The resolutions, duly signed, appeared in full detail on the front pages of the city papers. It satisfied the press and the consumers as far as farmers and the Dairymen's League were concerned.

At the next meeting of the full board the directors unanimously approved the Towner bill. Mr. Miller came back to my office alone the next day. He said that when he arrived in the city he found the League officials were "sitting on the lid of a volcano and didn't know it," that he was a "lawyer in a small town" and thought he knew the business of that community but in New York he was "all at sea." But he knew I would know what to do. He said he had a serious talk with the "boys" and "cautioned them that they were not yet out of the woods" and could "not afford to throw away any trumps," that Mr. Dillon and the Department had the confidence of farmers and he wanted them to "take Mr. Dillon into their confidence so that he could be in a position to advise and help them."

I asked what their reaction had been to that advice and he replied, "They said that 'the dealers do not like Dillon,' " and he had told them "the trick to divide the enemy and lick them separately was invented before the dealers were born."

The effect of the two published resolutions not only quieted the milk agitation and prevented a complaint to the District Attorney, but had two other results:

1. The dealers knew that through their influence the League Committee had been cool to the Towner bill and weaned away from the Department. In the October fight farmers had had the support of the press and the friendship of consumers. Now the city public and press had become suspicious of the organization and of farmers' good faith. Accordingly the dealers could be "stiff" about the new prices to farmers. The published resolutions, however, had restored confidence in the farmers and their organization. The

dealers and their Milk Conference Board "limbered up" and an agreement for the next six months was readily negotiated;
2. John D. Miller was retained as attorney for the Dairymen's League. The amount of the retainer has never been known publicly, but it was described as "handsome." In his previous interview with me, Mr. Miller agreed that it was unfortunate that a successful business man with executive ability had not been elected president of the Dairymen's League. When I met him after his retainer, he told me he had changed his mind and felt now that "Mr. Cooper was a good man for the place."

City Price Up Again.

In the second week of July, 1917, Borden's announced the third raise in ten months of one cent a quart to consumers. The price to consumers then was twelve cents for Grade B, Grade A fourteen cents. Cream and condensed milk were proportionately higher. This was the sequel to the defeat of the Towner bill.

Dealers now feared no opposition in the city market. The extra profits would make the cost of defeating the Towner bill look like thirteen cents. The farmer was then getting $2.00 a cwt. or $40.00 a ton. Feed had gone up to $60.00 a ton, leaving the milk price 50 per cent below a profitable price.

I published figures to show that milk could then be sold through stores at a saving of three to five cents a quart on distribution costs alone, and suggested that since the Towner bill was defeated, the Dairymen's League should make the demonstration from the farm-owned plants in the State. Experienced farmers corroborated the estimates and repeated the appeal to the organization, but nothing was done.

Killed a Good Project.

Previous to this, a delegation of eight dairymen headed by a banker from Dutchess County came to my office and said they were ready to build and equip the city plant. They were Harlem Valley producers and naturally would like the plant accessible to their territory, but they voluntarily put their natural preferences aside, and left the choice of location to me. Mr. Price, the chairman, had visited me before and was enthusiastic about the city plant. I took this farm committee to the Executive Committee of the League because the plant would require co-operation of the League to supply the milk. The committee was lukewarm to Mr. Price's generosity. He told me in parting that the committee was clearly against the proposition, and since it controlled the milk, nothing could be done about it. I knew he was right.

Shortly after this incident, William Boyce Thompson, who had a rating of $50,000,000 and was said to be worth more, and who had become interested in my food distribution work, and friendly to me personally, told me he had been looking for an opportunity to use some money for the public good, and was ready to pay for the city milk plant that I wanted. He said as soon as I felt ready to go ahead he would supply the cost. I was never in a position to assume the responsibility during his lifetime because of League official opposition, and consequently the uncertainty of an uninterrupted supply of milk.

CHAPTER XVI

POLITICS RUN RIOT

Middlemen Win.

When Governor Charles S. Whitman was canvassing for reelection in the Fall of 1916, his secretaries mailed me extracts from speeches he made in farm communities. Clippings of these speeches were sent me by his attendants and with the approval from farmers in the localities. The Governor praised the work of the Department of Markets to the skies, and pointed to his friend, John Dillon, as the type of honest and capable men he retained in the State service without regard to partisan politics.

When he returned from up-State he told me he had found that I had legions of farm friends up-State, and that the Department of Markets was the "life of his administration." About two weeks after his re-election he made a statement in which he said that Commissioner Dillon had done wonders with the Market Department but it was too puny. He proposed that it should be greatly enlarged, properly financed and made a help to win the war.

He appointed a committee of five to study foods and markets, and to report on needed legislation. George W. Perkins was chairman, Senator Wicks was a member, and George W. Ward, counsel. The last two were chairman and counsel respectively of the Wicks Committee previously appointed to investigate milk. Mr. Perkins was also chairman of Mayor Mitchell's Food Supply Committee, which had been standing two years. Seth J. Lowell, Master of the State Grange, and Clifford S. Sims, vice-president of the Delaware and Hudson Railroad Company, were the other two members. All of these were combined in the Governor's Foods and Markets Committee made up of 21 members.

I was familiar with all this plan before it happened. I had many times discussed foods and markets with Mr. Perkins, who seemed to be earnest and sincere. When I was president of the New York State Agricultural Society, I invited him to speak at the annual meeting on January 18, 1916. He did so. In that speech he said the dealers "insulted his intelligence" when they proposed to make the Department of Markets a bureau in the Department of Agriculture because he knew as well as they did, that the real purpose was to put the Department of Markets on a shelf where it would be out of the sight of the people, and where it would not do anyone any good.

When Governor Whitman first took office in January, 1915, the Department of Foods and Markets was not yet a month old. I went to him and explained the circumstance of my appointment, and my desire to get back to my publishing business as soon as possible, that he would probably like to appoint a man for the place, and that I had come to offer him my resignation. My term was six years. He asked me to stay, and I consented with the provision that he consider my resignation in his hands any time he desired to make a change.

During the year 1917, I discussed with the Governor, and his confidential man, George Glynn, the advisability of appointing a deputy who would become familiar with the work, and qualify as Commissioner, allowing me to retire. At Mr. Glynn's suggestion I discussed the matter with George W. Perkins. After election, Mr. Perkins told me that the Governor wanted to develop the Department and adequately finance it, but in view of my desire to retire, the Governor was concerned about "manning it."

It was impossible to do real worthwhile work with the meagre appropriation so far made for it. With the new attitude of the Governor and Mr. Perkins' interest, the Department would be amply financed. If so, and efficiently managed, I knew the Department would flourish. The Governor wanted to know if I would stay long enough to familiarize a new man with the work. I said that I would not stay in the office after a new man was selected, but I would give him all the time he required, and I would personally, and through my paper, help and support the department and its management so long as it devoted itself to the purposes for which it was organized.

A Change of Heart.

In the meantime the Wicks Committee, which had held hearings in the country during the summer and taken some helpful testimony, resumed its hearings in New York City. Here its attitude had changed. Up-State its testimony showed that farmers were producing milk at a loss, that the standard price of milk to farmers was fixed by Borden's twice a year for six months in advance, that in effect a trust existed in New York to fix prices for farmers' milk, that the dealers had shipped quantities of cream, and also shipped practically as much fluid milk as they bought, showing that cream had been extracted and that the skim milk had been mixed and shipped with the fluid milk.

Testimony also revealed that grain dealers gouged buyers of feed in the weights of grains for feed grinding. I learned by the testimony of the secretary of a middleman's association that, by direct lobbying with the legislators, he had stopped any appropriation for the bureau in the Agricultural Department for encouraging co-operative associations of farmers.

In New York, Counsellor Ward led his dealer witnesses to make the excuse that prices had been made in New York by custom like street car fares, and the dealers just innocently paid farmers what was left after they took their abundant share. Dealer witnesses were handled with silk gloves. Milk dealers not affiliated with the trust were led into technical traps, and those prepared to tell some unpleasant facts were deftly led away from their subject.

The atmosphere at the time was not just right to risk an open attack on the Department of Markets. Hence, Mr. Ward kept an outward appearance of friendship for it, but lost no opportunity to work in discrediting testimony, and magnified it in his reports. With Senator Wicks, Mr. Ward visited the wharfs, the produce markets and the milk plants. He found that New York City had the most efficient system in the world. There were, he admitted, a few small milk dealers who discredited themselves and caused criticism of all, but the Borden Company was little, if anything, short of perfection for efficiency and economy in distribution.

A Ripper Bill.

When the Foods and Markets Committee began to hold joint sessions in the late Fall of 1916, Commissioner Perkins invited me to sit in with them. By that time I had caught something of the trend of the Wicks Committee. The chairman and counsellor were both political wards of Senator Elon Brown, with whose policies I was familiar. I was not a member of the committee. I would have no vote. If Mr. Perkins wanted me as a member, the Governor would have made the appointment. I declined the invitation. The request, however, was repeated several times, so finally I consented to join them one afternoon.

Mr. Ward suggested that the secretary read what they had agreed upon so far as a background for my information. After the secretary had read several paragraphs, I politely interrupted to say that it was not necessary to take up the time of the committee to read more. I saw the trend of their minds, and I was sure that I could be of no use to them, and asked to be excused.

As I rose to go, Mr. Ward pleaded very innocently and appealingly to know what, if anything, was in their recommendations that I did not like. He entreated me to stay and assured me that he would not recommend any legislation that I opposed. I told him that it was not necessary for the committee to prepare a ripper bill in that case, that Chairman Perkins knew my resignation had been in the Governor's hands for nearly two years, and that I did not consider it a compliment to my intelligence to expect me to sit with a committee and approve a bill, to put myself in a false and humiliating position before the Legislature and the people of the State; that if any of them had the authority to accept the resignation, he could say so and they could get rid of me then and there; and that if not, the Governor had, and no ripper bill was needed. It was clear enough to me then that they wanted me to help draft the legislation so that as a party to it, I could not effectively oppose it whether I liked it or not.

Mr. Perkins accompanied me out of the room. Outside the door, he said I did right to speak out to "those fellows," and that he would seat me next to Governor Whitman at the dinner that night, and he wanted me to repeat every word of it to him. The dinner had been previously arranged. My seat was next to the Governor, but I did not need to repeat to him what I had said to the committee some hours before. He had heard it all right. I did tell the Governor that it did not look as if I could go along with his committee. He said he hoped I could, and invited me to come to Albany and talk it over with him.

The Wicks Bill.

If the report of the joint committee was otherwise fairly satisfactory, I was prepared to approve it even with the "ripper" provisions which to me had become academic as long as I had no part in the deliberations of the committee. I would have overlooked everything short of fundamentals in my desire to further the interests of the Department of Markets.

The report of the joint committee to the Governor on January 4, 1917, was harmless enough. It contained no new information. Its general statements might have been written by a rewrite clerk from reports of the Department and other published agricultural literature, but its conclusions might mean anything. It had the characteristics of a public feeler. The public statement that followed from Senator Wicks and Counsellor Ward made the purpose more evident. There were strong intimations that Senator Wicks and Chairman Perkins were not in accord.

I had for years stated definitely what I believed to be the fundamental requirements of a system for the economic distribution of food products. Mr. Perkins had publicly and privately approved them. He had not indicated any change of mind. Senator Wicks and Counsellor Ward were clearly at variance with these principles, but some one inspired a city paper to say that "Commissioner Dillon had agreed with Mr. Perkins, while up to that time at least, the facts were that Mr. Perkins had agreed with Dillon."

The Wicks Committee bill was released late in February, 1917. It was characterized as the "craziest piece of political patchwork" that had ever made an appearance in Albany. It consisted of 197 pages and was largely a useless readjustment of the agricultural law. It proposed to combine the Agricultural Department, Foods and Markets, and Weights and Measures in one department under a commission of seven members, which would also take over a part of the functions of the Department of Health. The new department would have an attorney and a secretary, whose annual salaries would amount to $54,000. The Commissioners of Agriculture and of Foods and Markets, would be appointed by the commission, but the term of the present Commissioner of Foods and Markets was continued four years beyond the terms of the old law, thus for the present time at least, eliminating the "ripper" provision of the bill as to me.

The City of New York was to be divided into milk zones, in each of which a single dealer would be given a monopoly of the distribution. The Department of Foods and Markets was to be subjected to painless dentistry. Every tooth in it was to be extracted. It could do none of the important things that it was originally organized to do and had been doing with the approval and commendation of producers and consumers of the state, with exceptions so rare as to be practically negligible. It was a proposed complete surrender of farmers' rights and interests to speculative and monopolistic combinations. As a concession to consumers, housewives and school girls were to be given instructions in the economic use of foods, and farmers were to have more agricultural schools, and "every resource should be employed by the State to increase the farmer's efficiency through increased intelligence concerning his profession." The profits of produce dealers and milk dealers were specifically safeguarded.

At the legislative hearing on the Wicks' bill in Albany, Counsellor Ward threw off his former disguise in his claims for the bill. Whether purposely or unconsciously he made his appeal on the ground that milk dealers must be protected in their profits, and that farmers must be restrained from obliging the dealers to "pay $2,000,000 a year for the dirt in their milk." He sneered at the suggestion that it was possible to pay farmers more and to charge consumers less. He charged that the feat was impossible and the claim insincere. He characterized anyone who opposed the Wicks' bill as a faker or a crook. The system of distribution in New York City was the best in the world and that included the distribution of milk by the Borden Company.

A short notice had been given farmers for the hearing, but they had an analysis of the bill through *The Rural New-Yorker* for three weeks and were alert. They filled the Assembly Chamber and were ready for the debate. They knocked out a proponent of the bill in every round, and no farmer spoke a single word for it, nor anyone else except those who wrote it.

George W. Perkins forgot the dealers' insults to his intelligence which he resented a year before in the same building at the Agricultural Society meeting, and actually joined the middlemen in an appeal to put the Department of Foods and Markets out of sight in the Agricultural Department. He departed from the fundamental principles that he had formerly avowed to say that farmers should not have authority to name the price for their milk or other products. Quick as light a young, clear-voiced, well-spoken St. Lawrence County farmer shot back this question: "Who should fix the price for the plow and harvester I just bought from the company that pays you a dividend out of its profits?" After that, Mr. Perkins was like the Irishman after a fall; he was not dead but "speechless."

The answer to Mr. Ward was even more decisive. He was told he had proved too much. He had just spent $25,000 of public money to prove that our marketing system was the best in the world and needed no improvement, and that the distribution of milk was little, if any, short of perfection. It was apparent that the food distributors who had been so unjustly maligned, were worthy of a place on the calendar of saints. But if this were all true, why had Mr. Ward asked the Legislature for $32,000 more to continue the investigation? What was the need of a massive Wicks' bill? Since he had proved so much perfection, the logical thing would seem to be for Judge Ward and his associates to clear up their desks, put on their coats and go home!

The Temporary President of the Senate, Elon Brown, tarried long enough to tell a group of farmers that "Senator Wicks had worked up the bill," that he, himself, had "not given it any attention," but he saw that "farmers did not want the bill and that's all there is to it." The Wicks' bill was dead. Farmers said then and there they did not want any bill in its place. "Leave us alone" they said.

George W. Ward left the Wicks' Committee.

The Fight Renewed.

The next morning Governor Whitman had a large group of farmers in his office. He told the farmers that some of them had said they wanted to take the Agricultural Department out of politics, and if they yet wanted to do so, then was the time. "But," he said, "gentlemen, you know the agricultural business; I do not know it. So, gentlemen, if you want this legislation, you must do the work yourselves or send me capable men to do it. You know them and I do not know them."

There were some farmers and some politicians in that conference, who knew, as well as I did, that the Governor was committed to the New York food distributors and to George W. Perkins for political and financial support in the late election, and that the purpose was to shape legislation along the line of the Wicks' bill to liquidate the obligation. Of course, that meant first and foremost to sterilize the Department of Foods and Markets. The dealers alone had contributed more than $300,000 in three years to that objective. But some farmers present actually believed that they were going to see politics taken out of the Agricultural Department. It was, but the politics were taken over to the Executive Department.

That afternoon, Senator Elon Brown held another conference of men invited by him in his office at the Capitol to consider new agricultural legislation. I sat and heard it all. Near the end, Senator Brown said he would like to have my views on the subject. I told him that I was opposed to it, that farmers at the hearing the day before had requested that no new farm legislation be passed that year, and that I believed they had spoken for their own best interests.

My office was in New York. Some days later I had a telephone call from Senator Wicks asking me to come to Albany and sit in with some of them to discuss a new farm bill. I reminded him of what I told Senator Brown on the subject, and asked him to excuse me. About a week later I had another call from him. He said a new bill would be written, and as it affected my Department I should be there to make suggestions. He asked why I objected, and I said, "From what has already transpired, Senator, I am suspicious of Albany's intentions." He then said they were all "white men" up there, and friends of mine. They were to have a joint meeting of the Agricultural Committees of the Legislature and Senator Brown particularly wanted me to come.

Before I went to the meeting I made a memorandum of several provisions that would strengthen the Markets Law, and enlarge the service of the Department to farmers and consumers provided the Legislature would supply funds to make the service possible. Alleging that his right ear was defective, Senator Brown insisted that I take a seat next to him at his left side. I presented my suggestions and Senator Brown accepted every one of them, with comments "good," "fine," "I approve every word of that," and so on. Then he said, "Just let me take those papers and I'll embody them in the bill."

A Shameless Farm Tragedy.

When the bill was ready Senator Brown had a large group of picked men to hear the bill read. Editor Charles W. Burkett, who inspired Brill's stunt in the crisis of 1916, was in a front seat to tell Senator Brown's willing ears that farmers wanted the new Wicks' bill. The members of the Executive Committee of the Dairymen's League and Counsellor John D. Miller were there with whetted knives in their breasts to assassinate the Department of Foods and Markets that six months before had taken them out of admitted failure, distrust and discredit.

Chairman Cooper had said that he was tired of hearing farmers say that "Dillon and the department had won the milk fight"; that he wanted "credit for the League," meaning himself. Outwardly he approved the Towner bill when to do so served his pressing needs. I knew, from inside information, that he was secretly allied with the dealers who opposed it, and also that

Miller and he wanted the second Wicks' bill passed. But this was the first time they had ventured open approval of it. Editor Burkett was yet being denounced by farmers and farm organizations because of his harmful interference in the milk fight. His paper was being shunned, and he was receiving caustic letters, but he had printed pictures of the officers, whom he had once denounced as sycophants to Dillon and so he felt qualified to misrepresent the sentiments of farmers.

Senator Brown had made skillful use of every suggestion I had given him. They were all in the bill, but as innocent and helpless as an infant in swaddling clothes. Every provision was for something to be considered, discussed or investigated. Not a single one of them carried an authority or a mandate. A shameful farm tragedy was being developed, but it was staged as a political farce in the twenty million dollar Capitol under the auspices of the Governor of the State.

The Farms and Markets Law.

This legislation limped along until May. Farmers had lost interest in it. It proposed to abolish the Department of Foods and Markets and create an agricultural council of ten men. The Commissioner of the Department of Markets of the City of New York was to be a member ex-officio, and the nine other members were to be appointed for ten years, not by the farmers who "knew agriculture," but by the Governor who "did not know it," and their successors to be appointed by joint vote of the Legislature. The council would administer the Agricultural Law, the Foods and Markets Law and Weights and Measures Law. It had come to be known as the Farms and Markets bill.

After many conferences, the bill was scheduled to pass, but interest in it lagged. Deals had to be made with the Democratic organization for votes enough to pass it. After it passed the Senate it was held up in the Assembly and it finally passed with the understanding that George W. Perkins was not to be at the head of it. For that reason Governor Whitman did not want it. He put off signing it and proposed a new bill.

Brown Bill Defeated.

Two days before a recess, an emergency bill with many features of the original Wicks bill was proposed with an appropriation of $1,500,000. It was thought that it would slip through, when the legislators came back after a short recess, before opposition could come from the country, but short as the time was, lively opposition did flow from the farms and the bill was defeated by an overwhelming majority. It was introduced by Senate Leader Elon Brown and bore his name.

The three-headed Farms and Markets bill was then signed by the Governor, but he deferred the appointment of commissioners. It destroyed laws that were helpful and provided nothing to take their place. Milk was practically left to the mercy of the trust. The dealers at the time were getting two cents a quart more for delivering milk than the year before. Farmers were getting the one-cent more which they had gained in the October fight, but because of increased costs of production due to the war, they were not as well paid as before.

When farmers had killed the first Wicks bill, they had won the season's fight. But when the Farms and Markets bill appeared, some of the leaders weakened. The enemy succeeded in the old-time trick of dividing the opposition forces. It made unimportant concessions, and then forced its bill through in spite of opposition. It stripped the Department of Foods and Markets of the power it had the year before to put the strong arm of the State on the side of farmers. This was done with the consent and connivance of the men who above all others might have been expected to protect it.

State Food Commission.

The principal purpose of the Farms and Markets Law was to pay the Governor's political debt to George W. Perkins. If the emergency bill had passed, it would have furnished a place for Mr. Perkins, so the Legislature adjourned, and the Governor began to frame a food bill for Mr. Perkins' benefit. In August he called a special session of the Legislature to pass it. This new bill created a State Food Commission of three members appointed by the Governor, who also designated the president of the commission.

The law provided that the commissioner could hold two State offices in order to permit George W. Perkins to accept a place in the commission without resigning one already held. The law would terminate at the close of the World War. It was a clumsy piece of legislation, but it passed the Senate with three negative votes. In the Assembly, 99 voted for it and 31 against. It was opposed by resolution of ten farm organizations and by farmers generally.

Governor Whitman appointed George W. Perkins to administer the new law. Farmers vigorously opposed the appointment, and the Senate refused to confirm it. Mr. Perkins went into the country to convince farmers of his qualifications. In a dispute he offered to pay for advertising to show reasons for farm opposition. The challenge was accepted at a cost estimated to be between thirty and forty thousand dollars. Then the Governor nominated Mr. Perkins again. Farmers continued their opposition. The Senate met again and in response to the wishes of farmers and city consumers, again rejected the nomination. The Governor then appointed commissioners, none of whom represented any agricultural interests of the State.

Farm Council Named.

It was then November. The Farms and Markets Law had remained dormant since its passage in May. The Governor then appointed the nine members of the farm council. The food dealers had at least three members, the city was represented, and the political exigencies of the Governor were represented. It was a farm council with no one to represent farmers. While it was proposed originally to take the "Agricultural Department out of politics," the Governor insisted that it authorize him to appoint the first commissioners. The men who attended the Governor's conference the morning after the defeat of the first Wicks bill, and were told they must take the places or provide other farmers for the place, were not invited to come back. Of the lists of names they sent to the Governor with recommendations for appointment, none appeared on the list of councilmen.

Farm Interests Demoralized.

By mid-winter of 1917-18 agricultural matters were in a state of chaos in Albany. The Agricultural Department had become completely demoralized. The Commissioner of Agriculture had been reduced to the position of errand boy by George Glynn and Secretary Orr of the Governor's staff. The Governor withdrew the appointments to the Farms and Markets Council. Everything was in a state of confusion and humiliation. Farmers had lost all confidence in Governor Whitman. I sat in a farm meeting in Albany and heard the chairman say that the Governor did not have a friend in the room. No one disputed him. He spoke what everyone in the room felt to be true.

The Governor did not treat farmers with candor. They resented his broken promises. He promised to eliminate politics from the Agricultural Department, but later he insisted on appointing the first members of the council and turned the department and his Food Commission over to professional politicians.

A large farm meeting held under the auspices of the Farm Bureau had passed resolutions demanding the repeal of the Food Commission law and of the Farms and Markets law. In Albany farmers stood by those resolutions. There were men in positions and others looking for jobs who were eager to keep in favor with Governor Whitman, but the strong, independent farmers would have nothing to do with his administration. Some of them said they would not accept places on the Farm Council, if appointed, though later when appointments to fill vacancies were made by the Legislature some good men accepted appointments on it.

CHAPTER XVII

THE FEDERAL MILK COMMITTEE

Plan To Divide Farmers.

The second six months' contract between farmers and dealers expired on September 30, 1917. In the meantime war conditions had driven milk prices sky-high. The demand for condensed milk for the armies was unlimited. In Wisconsin the condenseries were paying as high as $3.75 a hundredweight for milk. The price to New York farmers had averaged only $2.49 for 1917. The price to consumers had been increased three cents a quart, or $1.41 a hundredweight, within the year. The League Committee named a price of $3.10 for October, $3.34 for November, and announced that the dealers had at first objected but finally accepted. Dealers said there was no such agreement.

The Borden Company insisted that its agreement was to pay the full price only to the producers who were stockholders in the Dairymen's League, Inc. for Borden's fluid requirements. The price of the milk handled by other dealers was up to the League. The major part of the producers were not stockholders. It was a flagrant attempt to divide the farmers.

On pay day Borden's sent checks for the full amount to League stockholders but made deductions from the returns to other patrons for what they claimed to be their loss on surplus. The League Committee must have furnished the names of stockholders in accordance with the agreement. Farmers wrote me asking why they had been discriminated against, reminding me that I had told them their word was all I wanted and that they had fully co-operated with me in the fight just one year before. They had assumed that I was yet negotiating the prices.

When Borden's first announced the terms of the contract, I attempted three different times to show the Executive Committee that this policy would split the organization into two groups, that we had worked forty years to get dairy farmers united in one body, and that if the dealers were permitted to continue such policies we would soon be back where we started. But the thing was already done. I could not convince them, and they could not change it if they wanted to do so.

Many farmers who received the short checks from Borden's returned the checks and demanded the full price. Some advised their lawyers to collect the full amount. My understanding was that after receiving such letters and consulting their own attorneys, all such bills were paid in full. Who paid the difference was not revealed, but if Borden's had any surplus, it was worth more for condensing than the dealers paid for the fluid milk.

A State of Chaos.

It would be difficult to picture the state of chaos that existed in the milk business in the fall of 1917 and on into the following year. The dealers dominated the situation and made the most of their opportunity. The Borden Company reported that it closed one hundred plants. It claimed that previously it had taken milk from ten thousand farmers but in October had reduced the number to eight thousand. Many country receiving plants had no buyers for their milk and individual producers were obliged to make butter or cheese at home. Some dealers refused to pay the full agreed price and offered less.

There was gouging on the fat test, and the fat in excess of 3 per cent was not always paid for. Others paid the Grade C price or less. One farmer complained that his dealer paid him the Grade C price for his Grade B milk. This was a loss of ten cents a cwt. I gave this dealer a choice of paying the farmer the Grade B price for the eleven months or prosecution for cheating a farmer or for violation of the health laws. He paid $150. I figured a shortage for a Delaware County farmer amounting to $21 and collected it for him, but few farmers made definite complaints. The League had no local organizations to correct these abuses. The Department of Markets had been converted into a bureau which made no attempt to check these abuses.

Previous to this time stores were allowed to buy pasteurized milk in cans and bottle it under hygienic regulations, but in the fall of 1917 the Board of Health passed a regulation requiring that milk must be bottled where it was pasteurized. Then the dealers charged storekeepers for bottled milk the same price they charged for home delivery. This rule favored the dealers and limited the stores' service and business.

League officials organized the Co-operative Milk Marketing Association to sell the milk of farm-owned plants to grocery stores. The association made little progress because it operated under the milk trust rules. Later it was announced that two plants, one in New York City and another in Brooklyn, had been purchased from the Modern Dairy Company to sell milk direct to stores at nine cents a quart, the stores to sell at ten cents a quart to families. At the time the stores were selling at twelve cents a quart. But the plan was never put into effect. The dealers could probably tell why.

R. D. Cooper, President of the League, went to Washington and invited the participation of Herbert Hoover, the Federal Food Administrator, in New York milk affairs. The Borden Company was in close touch with the Federal Food Administrator. Mr. Hoover appointed a Federal Milk Committee for New York. In a conference with the Federal Food Administrator, the Executive Committee of the League agreed to reduce the October price eight cents a hundredweight, to $3.02, and at the same time agreed to allow the dealers to charge consumers an increase of seventy cents up to $8.28 per hundredweight. The spread was then $5.26.

Later, in a council of Federal Food Administrators, Borden's insisted that because the farm price advanced in November it would be necessary to add another cent to the consumer. A request was made to ask that the League suspend the increase in the November price. No one suggested a reduction in the cost of distribution in the city, though the city price was then fourteen cents for Grade B, and up to twenty cents for Grade A, for which the farmer got one-third of a cent a quart above Grade B.

The dealers and the food administrators outwitted the League

Executive Committee. Chairman Cooper yielded to their demands and called the directors together to consider the reduction of price for November. The board promptly and properly rejected the proposal with a healthy bang. Then Mr. Cooper yielded to a second demand to call the board of directors to reduce the price for December, 1917, a half-cent below the November price and to agree with the dealers to accept a price to be fixed by the Federal Milk Committee, for the following three months. After a lively debate the board divided on this proposal, but the majority finally sided with President Cooper, voted to accept the December reduction and to accept the price fixed by the Federal Milk Committee for three months. There was not a person on this new committee who depended on the returns for milk for his business or his living, and only one who ever owned a cow. The milk dealers made no concession whatever.

By this time John D. Miller, the official counsel, had gained domination over Cooper. He had become dictator. Cooper took the criticism but "Boss" Miller was making friends with wealth and power.

Dealers Consistently Favored.

The Federal Milk Committee agreed to fix prices on the cost of production plus a reasonable profit, and if it appeared that the December price was below that standard it would be made up by the added prices for subsequent months. After its long investigation it fixed the January, 1918, price to farmers at $3.52 per cwt. This was an increase of fourteen cents per cwt. At the same time the dealers got an increase of forty-seven cents. The Federal Milk Committee reduced the February price to producers to $3.34, but made an additional allowance of twelve and one-half cents to make up for the under price for December. We will see later that Sheffield Farms' profit for that year was $774,000.

The committee acknowledged that the price to producers did not cover cost of production and profit, which it had agreed to allow after investigation. It admitted that it made no investigation of the dealers' charges for distribution. It also admitted that the dealers' system was expensive and wasteful but it pleaded that the committee had no power to reform the system. This assumed limitation of its powers and duties was not convincing.

The committee ascertained the cost of production of milk and promised to add a reasonable profit to this in the price to farmers. It did not do so. It accepted the book cost of distribution shown by the dealers and added a profit. It allowed the dealers their "wasteful and extravagant" costs but chiseled farmers on production costs found by itself on its own deliberate investigations. The committee continued to fix prices during the year 1918, except for the month of June. Its discrimination in favor of the distributors was consistent during the whole period of its existence. The committee set the May price at $2.45. This the League officials approved, but at the same time agreed with Borden's to accept $1.80 for June. The Federal Committee alleged that the League's concession for June was an embarrassment to it, and that to name the same price would make the committee look like a "rubber stamp."

A Proposed Surrender.

During the month of May, 1917, R. D. Cooper gave out what seemed to be a feeler, to the effect that League officials and the Borden Company were negotiating for future sales of Borden's supply of milk, involving a surplus provision and an option by the League to buy the Borden country plants. During the month following other references to the matter appeared in local country papers, but details were not available. In the month of June C. A. Weiant, president of Borden's Farm Products Company, gave me an interview. I sent him a proof of my report of the interview which he approved. He said he would like to send it to Mr. Cooper. This he did with my consent, and returned it as approved by both parties. It was printed in the June 15, 1918 issue of *The Rural New-Yorker* as follows:

"Borden's Farm Products Company has agreed to buy the milk needed for its New York fluid trade for the months of May and June from the Dairymen's League insofar as the League could supply it, with the understanding that the League would take care of the surplus, the price to be $2.34 in May and $1.80 in June in the 150-mile zone for 3 per cent milk and 4 cents extra per hundred for each one-tenth of one per cent butter fat. The Borden Company is to take its requirements of fluid milk for its city consumers, including cream and buttermilk. The surplus is a problem for the League. The Borden's Farm Products Company to take all the milk and sell it to any condensed milk company, including the Borden's Condensed Milk Company, or any manufacturing company, or they will manufacture it into butter or cheese. The loss sustained on the re-sale or manufacture of the surplus is to be deducted from the account of farmers who sell milk to the Borden's Farm Products Company, but who are not League members. The losses due to the handling of the surplus will be pro-rated among the non-League members on the basis of 100-pound lots and deducted from the returns of the non-League members. If it works, this plan would give League members the League price in full and put the losses due to the surplus on the non-League members. Producers have been notified that they will get full returns for May, because the company is not able to figure the surplus loss in time for June 15th payments, but the May losses will be deducted from the June bills payable July 15th, and the losses for July will be deducted from the August 15th bills.

The Borden's Company will use its discretion in the distribution of the surplus, re-selling or manufacturing the milk as it finds for best results. If non-members refused to sell on this basis, or if all milk producers became members of the League, or if the entire amount of milk supplied by non-members be not enough to cover the surplus loss, the League would be obliged to find some other method of adjusting the loss. The problem of taking care of the surplus is entirely up to the League. If the plan that they have suggested, to take the loss out of the non-members, is not practical, then the League shall be obliged to find some other way to make good the loss to the Borden Company. By July 15 the League will owe the Borden's Company for the surplus losses in May, and if the company cannot recover this loss from the non-League members, it shall then put it up to the League officials and it will be their duty to find a way to make up the loss to the Borden Company, and to find a means to take care of it for the future.

The Borden's Condensed Milk Company is not a party to the contract, and is not a party to the contract with Borden's Farm Products Company.

There is no formal contract for May and June, but there are written memoranda to supplement verbal understandings.

The Borden's Farm Products Company has also agreed to sell its country milk plants to the Dairymen's League at investment values, enabling the company to get its money out of the land and equipment. There is no formal contract for this option, but the verbal agreement is supplemented by written memoranda. The proposition is to sell all the plants at one time at investment values, but the Borden's Company has indicated its willingness to wait a reasonable time for cash to allow the League to accumulate funds. The proposition is being considered to allow the purchase of the plants on installments. On this plan it is proposed that the League pay in regular installments for a number of years. The Borden's Company would continue to own and operate the plants until the sale price is paid. Nothing definite has been settled, however, but it is expected definite statements can be made in about two weeks."

The same issue contained the following editorial:

"We can hardly credit the belief that League officials are innocent enough to think that the surplus loss can be charged to non-members. Evidently the Borden's officials have no such illusions. Since the Borden's Company has purchased its supply from the League, it will buy of no other producer while the League has a supply. Consequently they cannot charge non-producers with anything. If there is a loss on account of surplus in May and June, League members will pay it. Any other theory is food for babies.

The option proposition is equally romantic. Farmers have no notion of buying old plants at cost, and leaving them in the hands of the original owners for a number of years. To accept such a proposition would be to put their necks in the yoke for the rest of their lives. Farmers will own their own plants, and establish a reasonable cost for distribution, so that they can increase consumption of milk by getting it to the consumer at a reasonable cost. They will make no contract that will hamper them in that purpose.

Stripped of romance the situation is that the Borden Company pays $2.46 for May and $1.80 for June milk, or about four cents per quart for June. They sell bottled milk for 13 to 20 cents per quart. If they sell one and a half millions in June as estimated, they will get four million dollars for delivery aside from the original cost of the milk—some harvest. The surplus they will sell to condenseries, or make it into butter or cheese. They will sell the butter and cheese, and take out the cost of making, selling and storing. The producer will get what is left. Romance about it all you will, such are the facts. The non-member romance and the option allurement may attract the minds of League members from the June prices, but the propositions will not flatter their intelligence.

The officers of the League who deposed President Brill in 1916 now owe him reparation and public apology. His proposed terms with the dealers included less than one-half the concessions in the present surrender."

THE FEDERAL MILK COMMITTEE

It was further understood later on that the installments would extend for 25 years and that the Borden Company would control the plants and the milk in the meantime. As soon as the farmers had made substantial payments they would be helpless. They would have no other market for their milk, and if they quit before maturity of the contract, they would lose all. Borden's would have full book price for its junk, retain ownership and use of the plants, and a 25-year option on the milk at their own price and terms. In 25 years the plants would be obsolete and the farmers would have no money to replace them. But after the scheme was fully exposed in *The Rural New-Yorker* the farmers' verdict was emphatically "No," and that ended it. But we will see it again in another form.

CHAPTER XVIII

THE COUNTRY MILK COMPANY

An Appeal For Help.

Mention has already been made to a reference in R. D. Cooper's first letter to me on June 4, 1916, concerning a purpose to organize a number of farm-owned plants into a company to distribute milk in New York City. The Department of Foods and Markets had proposed a plan to establish a city plant to demonstrate the cost of city distribution by a State-wide organization under the auspices of the State Department. It was also proposed that the plant and its operations would develop a stable market for milk and other dairy products. Mr. Cooper proposed that the project be undertaken by a small group of producers under the management of his Executive Committee of the then dormant League. It would put him in the city milk business.

The purpose was repugnant to co-operative principles, which insist on unity and equal opportunities and benefits to all members of co-operative associations, but the violation of these fundamental principles persisted in the League management. No one but the officials themselves knew whether or not their ambition to get into the city milk business was responsible for their violation of sound co-operative and business principles. The disaster to milk producers was the same in either case. It was a thin-edged wedge driven into the solid phalanx of the League.

In their first sale of milk after the repeal of the law which created the Department, the League officials yielded to Borden's insistence for the privilege of buying what milk they wanted from whom they wished to buy, leaving all other producers without a market to shift for themselves. When, with League officials' consent, Borden's and the other dealers shunned the co-operative plants, and closed some of their own fluid plants, League officials then organized the Co-operative Milk Marketing Association.

THE COUNTRY MILK COMPANY

As already stated, this did not succeed, so the Country Milk Company was organized on a more pretentious scale, but under the same handicaps. The principal officers were the same as in the League. Mr. Cooper was then in the New York City milk business. It was admitted that this was a deliberate policy, and intended to increase official revenues. The new company gave Mr. Cooper and other officers extra salaries. Three of the Mutual McDermott routes were bought and a plant was rented on West 125th Street. N. A. Van Son, who had helped me during the 1916 price fight with dealers, was engaged as manager.

This was Mr. Cooper's fourth violation of the co-operative principle of unity, leading to a divided membership. The Country Milk Company was a city milk dealer, and like the Co-operative Milk Marketing Association it conformed to all the rules of the city milk trust. No other dealer would sell milk to the consumers on its three routes. It was not permitted to sell milk to consumers on other dealers' routes. It must charge consumers and stores the price fixed by the dealer's organization. Its trade was thus limited. Expansion was impossible and the volume of milk refused by the dealers and flowing to the company to sell was much more than the three routes could absorb. So limited and bound, it was a failure before it started.

One day in the latter part of May, 1918, Manager Van Son came to me and said, "they" were "losing their shirts." In other words they were fast losing money. He wanted to break away from the dealers and sell milk to stores at a price that would increase his outlet. He wanted to sell the stores at a low price so that the price to the consumer would be 8 cents or two quarts for 15 cents. The dealers' price was 11 cents. He said the producers authorized him to do so.

Mr. Cooper had fixed the price to farmers at $1.80 for June or a little less than 4 cents a quart. Van Son pleaded with me to help him. I was reluctant to mix in the League affairs, being again absorbed in my publishing business. The League had made a mess of selling milk. Cooper took up policies that I had advocated, but changed them to suit the purposes of the official management with a resultant loss of the membership, and failure because of his affiliations with the dealers, and his own selfishness and incompetence. Van Son, however, reminded me that I had advocated for years just what he was now ready to undertake, that it was the only thing that would save the League, which I was anxious to preserve; besides, when I needed help in the department he had helped me, and now he wanted my help, not for himself but for the dairy interests.

He would sell bottled milk to stores all over the city as well as loose milk in cans to be re-sold to consumers with their own containers. He did not want me to do any work or to go to the plant. What he wanted was my name to win the confidence of the stores and the public, and especially of the city press, which was essential to a quick success. I figured that farmers could be paid the already fixed price of $1.80 a cwt. and all other costs covered, including storekeepers' profit, out of seven cents a quart paid by the consumer. I told Van Son that if he and the stores would work on that basis I would help him for one month, June, provided he would guarantee that he would go through with the proposition and not offer me any money for what I might be able to do. To this he agreed.

Success Killed the Plan.

I sent a statement to the newspapers. The reporters called together as they did in 1916, and we got stories with double heads at tops of columns on front pages. Some of them gave us editorial approval. It was a success from the start. The papers kept up the stories of "7 cent-milk" to consumers as the consumption increased until the output had doubled. More than a thousand stores were selling "7 cent-milk" in less than a week. One chain of 152 stores which had sold no milk before put up milk signs and sold "7 cent-milk." A single month had demonstrated that milk could be distributed in New York at a cost of three cents a quart. Even on a small scale, sales had doubled in three weeks. With the indicated increase of consumption and with proper equipment, it was quickly demonstrated that every quart of milk produced on New York farms could be sold and a reasonable profit realized.

Then the big dealers woke up and got busy. The old trick of stealing and dumping milk left at the store before the storekeeper arrived was renewed. Some "thugs" merely dropped junket tablets into the cans to sour the milk. But Mayor Hylan, Dr. Royal S. Copeland and Market Commissioner Day wrote me commending the work, and the District Attorney promptly put a stop to the interference with the milk. Some of the offenders were arrested. Then the dealers, big and little, held a meeting and threatened to influence drivers to strike and refuse to deliver the milk. But drivers' families use milk too, and they stuck to their jobs.

Then the dealers appeared in a body one morning at the Dairymen's League office and protested against the sale of "7-cent milk." Mr. Cooper put the blame on Van Son, so Van Son was called to the office. He stood up for the success of his plan, but the dealers turned again to Cooper, saying that he was the responsible head of the Country Milk Company and that they held him, not Van Son, responsible for the acts of the company, and unless he stopped the sale of "7-cent milk" they would repudiate their contracts with him for all the milk bought through the Dairymen's League.

That broke Cooper's resistance, so he told Van Son to stop. This Van Son refused to do on the ground that he had contracts with the stores for the full month, and also was pledged to me and to the press to carry on for the full month. Cooper forced Van Son to discontinue the sales after the 30th of June, and shortly thereafter Van Son resigned as manager of the Country Milk Company.

The local Federal Milk Committee again fixed the price for July. It was $2.25 per cwt. to producers. It also fixed the price to consumers. Whether or not Chairman Cooper had any influence in making this July price is a matter of speculation.

Two months later Mr. Cooper prepared a statement which he printed in the *Dairymen's League News* contending that Dillon had nothing to do with the sale of "7-cent milk," but claimed credit for what other men did. He asked other directors and Van Son to sign his statement before he published it; the directors refused and Van Son told Cooper he would not "sign his statement for a million dollars."

As a sequence to my attempt to help Van Son save Mr. Cooper's shirt as Van Son put it, Fred H. Sexauer recently developed a fictitious story that he likes to print at times in his official organ, *Dairymen's League News*. The story leaves the Country Milk Company, the Dairymen's League, the identical boards of directors of both corporations and the farm co-operatives entirely out of the picture. His story represents me in a private milk business with Van Son selling milk at a price that could net farmers only something like 13 cents a cwt., and actually never returning the farmers anything. One time he printed a muddled edition of the story over the initials of Roswell Cooper. One can imagine Mr. Sexauer chuckling over his smartness in hanging his dirty and odoriferous linen on my door knob, but to me it seems almost incredible that he should want to remind anyone of the perfidious record of the Country Milk Company. Of course, it is 20 years past, but there are many yet living who were the unfortunate victims of one of the League's most conspicuous disasters.

Ends In Bankruptcy Court.

The Country Milk Company was receiving the milk of 28 cooperatives. When N. A. Van Son, its manager, resigned he said that if he was permitted to sell the milk at less than the milk trust retail prices, he could dispose of all the milk, pay farmers the League price and make a profit, that he was not permitted to do so and in consequence a surplus was created to manufacture at a loss, and that there was useless duplication of official salaries and expenses. The 24 directors met monthly but never had a profit and loss statement. One half the milk bills were paid in the months of June and July. Then payments stopped entirely.

At the meetings directors asked for statements of the condition of the company to enable them to do their duty as trustees of producers. President R. D. Cooper refused, making the excuse that dealers would know their business. He admitted in the same meeting that he had restored the trust-prices to store consumers

THE COUNTRY MILK COMPANY

on July 1. In the following month John D. Miller, the League counsel, publicly defended the Country Milk Company and denied that Mr. Cooper had restored the trust-prices.

By that time dairymen, who had helped build up the Dairymen's League and wished to preserve it, attempted to get information as to its affairs. No statements were issued but members were told they could go to the office and get information. A meeting of dairymen appointed a committee to visit the office and get the promised information. In the morning the members of this committee were told to call at five o'clock, P.M., and they could have it. They called and were told they could not have it. They were offered a file of vouchers to examine but were told they would not be allowed to copy anything. They were simply refused the information, which had been publicly promised to all members.

An Appalling Breach of Trust.

The dairy farmers at Bullville, Orange County, had completed a co-operative plant costing about $32,000 and at President Cooper's solicitation had shipped the milk to the Country Milk Company to be paid for at the price fixed for all producers. In the early fall they came to me in great concern. Some of them had borrowed money at the bank on their own notes to put in the equipment and to pay producers. They owed money on the plant; they had received no money for their milk all summer and were yet unable to get even a partial remittance. Bills were due them for four months amounting to about $20,000. The bank notes were coming due and they feared they would lose their farms. They asked me to tell them what to do.

I first asked what the statement of the subsidiary showed. They had never seen a statement and did not know anything about its finances except that their bills were not paid. I then told them to go to the League office and demand information to justify them for continuing shipments. They appealed to me to get it for them. I could not refuse. They were my friends. I had encouraged them to build the plant.

I took their credentials to the city office the next morning and asked to see the records. The League official objected because the order had not been sworn to. I got the verifications and was then handed some incomplete statements. But it didn't matter. I saw that the concern had been bankrupt for months, while the farm credits were piling up and increasing month after month. A more deliberate swindle and breach of faith I had never known.

I wired the farmers to meet me at the plant. I made the report that night. Again they asked me what to do. I wrote out a telegram notifying the Country Milk Company that there would be no more shipments of milk and advised them to send it. They did. They never got a cent for their four months' shipments.

Finally the Addison County Co-operative Dairy Company of Middlebury, Vt., brought suit in the United States District Court for $13,000 against the Country Milk Company and the Co-operative Milk Marketing Association, R. D. Cooper, president of both organizations. In the month of March, 1919, the court appointed Edward J. McCrossin, a lawyer of 149 Broadway, New York City, as receiver. The complaint alleged that both concerns were mismanaged and owed farmers $400,000. The receiver-attorney said the concerns had been out of business several months, but accountants had been drawing $1,000 a month and H. J. Mosher, the treasurer, continued to draw at the rate of $6,000 a year.

The losses to farmers were estimated up to $500,000, but the exact amount was never known. The officials refused to give out the figures, and farmers' losses were so overwhelming I advised my friends to spend no money in litigation. It was my information at the time, however, that the officers received their salaries in full to the last penny. The complaint alleged that $50,000 had been paid out after the company had become hopelessly insolvent, but the receiver paid no dividends.

Not Co-operative.

I had objected to the organization of the Country Milk Company and other subsidiaries on the ground that the plan of the League promised equal benefits to every member, a united organization, and equal voice of every member in the management.

This subsidiary violated all these provisions. In the one experiment in the month of June, Van Son had demonstrated that milk could be sold three cents below the trust price at a profit after paying farmers the price fixed for them by the League itself, but the dealers had Cooper under control and stopped the one system of distribution that would reduce costs, increase consumption and benefit producers. I held, however, that except possibly for small demonstrations, the League could not expect successfully to mix wholesaling and retailing of milk in the same market, that it should confine itself to wholesaling milk and cream, and at the same time encourage and develop small dealers to buy and distribute its milk to families and stores supplying family needs.

The attitude of government agencies in price fixing for farmers was revealed in the fixing of milk prices for October, 1917. The cost of production for September, computed on the Warren formula, was $3.65 a cwt. The Federal Food Administrator and the Dairymen's League Executive Committee held a conference which agreed on a reduction of eight cents in the October price to farmers. It was explained that Herbert Hoover, then Federal Food Administrator, had requested that farmers accept this reduction to help win the war. But the same conference allowed the dealers an increase in the price to consumers of seventy cents a cwt. over and above its previous allowance of $7.58 per cwt. We published the statement and farmers called on us to explain how a reduction of eight cents to the farmer and an increase of seventy cents to distributors would help win the war. The incident shows that specious argument does not necessarily fool farmers.

The World War Ends.

On the morning of November 11, 1918, word reached America that Germany admitted defeat. The World War ended. Prices of all products and services were high. Butter, cheese and condensed milk were relatively higher during the war than fluid milk and cream. Farmers had particularly resented the small reduction in their price for October, 1917, not so much because of the eight cents per cwt., as because of the pretense that it was made to help win the war when at the same time the distributors were granted an increase of seventy cents.

Farmers More Alert.

For some time there had been general criticism of the League management, and demand for a change in the officers and executive committee. The collapse of the Country Milk Company and its attendant heavy loss to dairy farmers, as well as the scandal of it to the organization, the gouging on weights and tests, the losses through bankrupt dealers, and the chaos created by the dealers' refusal to take all the milk, convinced dairymen that they should themselves give more attention to the management of the League.

An election of directors and officers was due in early December. As head of the organization, criticism focused particularly on R. D. Cooper, but John D. Miller was still the boss. He wanted to preserve that status as it was. The price of November, 1918 milk was raised to $3.81 per cwt. That looked as if the Executive Committee had taken a strong stand and triumphed over the Federal Milk Committee and the dealers, but it was soon revealed that the price to consumers increased 47 cents a cwt. or one cent a quart. This made the city price 19 cents a quart for Grade A, 16 cents for Grade B.

At the annual meeting in December, 1918, the delegates took a more independent stand than formerly. The retiring directors were mostly re-elected. This continued the old officers and the executive committee, but fifty resolutions submitted reflected the abuses and unrest in the country districts. These resolutions were referred to a committee, which thought best to report only summaries of the most important resolutions.

In effect, this summary questioned the wisdom of canvassing new territory to increase the flow of milk to New York, requested an extension of the basic freight zone, urged that officers give their time to the League work and not accept other salaried positions, that all dealers be required to make prompt returns, at least monthly, to give patrons written data of their weights as delivered, that the law bonding dealers be strictly enforced and that all members receive equal benefits. These resolutions from the membership merely directed the management back to the original plan of the organization. They indicated a purpose on the part of farmers to give more attention to the work of the organization in the future. It was felt to be a healthful sign of progress.

CHAPTER XIX

THE 1919 MILK STRIKE

The Strike A Surprise.

In its proposal for a price for January, 1919, the Executive Committee followed the Warren formula, which showed the cost of production to be $4.01 per cwt. The dealers offered $3.60, and on the last day of December, 1918, the League officers sent out directions to farmers to ship no milk from January 1st on until further orders. That meant a strike. After several days Governor Smith responded to an appeal to intervene. He came to New York and after interviewing both sides appointed the members of the League Executive Committee and the bargaining committee of the dealers' Milk Conference Board as a committee to settle the dispute between themselves. It was rather a novel thing to do. It let the Governor out but it did not work.

Before it started, farmers had no intimation of even the possibility of a strike. They were not prepared for it in any way. The first days the farmers were in a state of chaos, with oceans of milk and no means to care for it. Much was dumped on the ground the first days. Then some found cheese factories to take it, others separated it and shipped cream to butter factories. Many devised means at home to make butter. But as in the Light Brigade, "Was there a man dismayed? Not though the soldier knew, Someone had blundered." Farmers remembered the past. Recent experiences were fresh in their minds. They recalled their triumph of 1916. They realized the importance of winning the fight on hand and they did their part in work and sacrifice. There were areas in which not a quart of milk was shipped during the period of the strike.

The Crisis In the Strike.

During the third week of the strike, John Arfman and Milton

THE 1919 MILK STRIKE

Lane, two prominent farmers of Orange County, active in dairy organization, came to my office in New York. Mr. Arfman said, "We are whipped." They came, he said, to consult as to what was best to do about it, and particularly to ask what I thought about breaching certain contracts that had been made to deliver milk over the strike period. He related that when the Country Milk Company gave up in November, there was a demand for milk and contracts were made for the milk of some farm-owned plants up to the following April. There had been no restriction on the buyers as to re-sale with the result that this milk was being diverted to the big dealers. The dealers were better organized than they had been in 1916, and better prepared. They were getting more milk from outside the State. In fact, they were getting 85 per cent of their normal supply.

I did not assume to advise as to the breach of contract, because I had not seen the written agreements, and while I knew of the sales, I did not know enough of the circumstances to know whether or not there were good reasons to justify a breach of the contract. I did say, however, that when I was a boy I had made a bad bargain and I was advised to refuse to keep it because I was then a minor, and not legally responsible. But my father told me to take my loss and pay up. I learned afterwards how sound and important his advice was, and I had never even considered a deliberate breach of word or contract. As a matter of fact, Arfman and Lane were high-class men. They felt about it as I did, but properly enough wanted to get the opinion of others on the subject.

After we had fully discussed the situation, we agreed fully on the causes that led to the failure of the League to repeat its original success, and on the cause of the failure of this strike. Because of the hopeless conditions, I advised that a small committee of the most capable dairymen be authorized to contact the dealers and make an agreement on the best terms possible for all the milk from the date of settlement up to March 31st. That would mean about seventy days. At the same time I advised them to see that no contracts were made for deliveries beyond that date, and make provisions to organize the association substantially on the plan unanimously adopted and approved by dairymen two years before, at least to the extent of preserving the fundamentals of co-operation.

To keep up the morale of the membership, I suggested that the price for the remaining days of January be kept as near as possible to the original asking price. Then with a united membership, all the farm plants available for use, and with full control of all the milk, the association would be on even ground with the dealers for negotiating further sales. If the dealers persisted in demands for a price below the cost of production, we could proceed with our reserve plan of marketing milk through the city stores.

A meeting was held in Utica two days later. I expected that the settlement of the strike would be arranged before the meeting and that the situation and the settlement would be announced at the meeting. What happened was that the farmers came full of fight and determination. They were resolved to hold out all winter. They had no comprehension of the situation. President Cooper had not kept them informed. He did not confide in them. On the contrary, he made a speech saying that he would fight "until hell froze over" before he would compromise or surrender. The next day or two a settlement was announced. The terms of sale were: Dealers would pay for the remaining days of January $4.01 per cwt., for February $3.54, for March $3.31. So far it followed my suggestion to Messrs. Arfman and Lane to the letter. The price for the eight to ten days of January gave us an opportunity to claim a victory, and the officers will tell you today that they won the 1919 strike.

No Protection For Producers.

While we all made the most of the strike settlement, claiming victory because of the price conceded for the few remaining days of January, farmers soon began to complain of the low prices in the agreement for February and March. Feed prices had gone up to $60 a ton. Wages and other supplies increased the cost of production. All of the old abuses and complaints continued. These included short weights, low fat tests, slow payments and inspection tyranny. Every day brought letters from dairy farmers. The

THE 1919 MILK STRIKE   153

following extracts will give an idea of the general trend of these complaints and protests:

"When the strike was settled, Borden's paid us 23 cents per hundred less than the League price for January, February and March, and short weights amount to $14.33. Will you take up this shortage both in prices and weights?

Vermont W. B. WHEELOCK."

"In April, 1919 the Pharsalia Creamery returned producers $2.69 for a hundred pounds of milk from the manufacture of butter and cheese and in May the same creamery returned $2.94 per hundred pounds of milk. This is better than the fluid milk prices I receive.

New York M. G. F."

"I sent a sample of my milk to the Experiment Station for analysis June 20 and got a test of 40 per cent. The cream company's test for the same day was 35 per cent. According to this they are taking nearly one-fourth of my butterfat. What can I do about it?

New York PRODUCER."

Real Co-operation Advised.

Since to all appearances nothing was being done to relieve existing conditions or to revise the League into a real co-operative farm organization, I tried to epitomize the ideas and sentiments of farmers in a short editorial in *The Rural New-Yorker* of March 22, 1919, as follows:

"We approve a plan to handle milk to the end that:

1. All members of the Dairymen's League be treated alike.
2. To incorporate local branches and regional organizations, and
3. To build and operate a number of manufacturing plants to furnish manufactured products, if need be, and to take care of surplus.

These features among others we formulated three years ago, and they were formally approved by dairymen then. But we do not think it wise to organize from the top downwards, so that the central organization should own and control local plants. On the contrary, we believe that the local branches should be incorporated and own their own plants, and collect and distribute the money for their own milk.

The local plants should be affiliated into a central organization to sell the milk and do the things that it could do better than the locals; but instead of making the local corporations subservient to the central organization it should be controlled and regulated by them. With one good expert creamery manager or engineer and a sure outlet for milk, local plants could be built and managed locally cheaper than from one centre, and it would be a simple matter to equip the best located of them to handle surplus. This plan would be complete with a city plant to encourage and supply small dealers and to demonstrate a cheaper cost of distribution through stores.

This business plan has been held in check for three years. It requires little centralized capital and little or no centralized operation. It must come sooner or later. Why not put it in operation now and avoid the possibility of disaster or continued losses in further delay?"

Farm Control Stressed.

Further to stimulate thought and discussion on the subject, *The Rural New-Yorker* offered $150 for the three best plans of organization for New York milk producers. This was prepared before the League officials made the March 22nd statement, and published the same week. The winner of the first prize of $100 was George C. Porter, of Upper Lisle, N. Y., a country-bred young man and a college graduate with a clear head. The winner of the second prize of $35 was John Anderson, of Morrisville, N. Y., a practical farmer and a man of practical experience in farm organization work. The third prize of $15 went to P. H. Norberg, of Bloomville, N. Y., who had experience in Denmark, where farm co-operation has had its greatest success. Many others sent plans, and it was difficult to decide which was the best.

All of them without exception insisted that the machinery of organization should be devised so that farmers could control and operate the business for themselves. They insisted that the organization must begin with the farmer on his farm and be developed through local associations to a central body, which at all times would be directed and controlled by farmers through their local associations.

They decried the tendency of self-appointed leaders to put all power in a centralized corporation and clothe it with authority over both the producers and their local associations. All the proposed plans provided for full information for the members, and a full detailed and accurate accounting of milk and money. All insisted on efficiency and economy in the management that the price might be made attractive to consumers so that consumption might be increased. They stressed the need of plants in desirable locations and their equipment to manufacture surplus milk and also to make a ready market for farmers who were not equipped to comply with the regulations for the fluid milk markets.

A Prize Plan.

The following is a condensed statement of one of the prize plans. It is typical of practically all the proposals:

"The Dairymen's League is a corporation like the ones that broke up the early Danish farm organizations. The members have no means of expressing their choice for officers or measures. At Utica not more than one-third voted for the new plan but it was called unanimous. A lawyer is the president. This is not co-operation. Farmers did not plan it or run it. They do not know yet what they lost in it. To call such management co-operation is unfair."

"My plan is for a real co-operative organization. Make a definite plan through a committee of the best members and experts in the business. Adopt a fixed policy and stick to it. Organize local plants where needed and affiliate these units in a State body."

"The officer must not be allowed to make by-laws, but reserve this power for the members exclusively."

"A good business manager should be appointed to conduct the business."

"The central body should be the sole agency for the sale of all the milk of members. There should be no subsidiaries and only one organization. True co-operation is economic."

"The members' contract should be with the local units. These local units should control and direct the affiliated State body."

"To vote intelligently the members must have full and accurate information. A full report of all meetings must be kept, and members have free hand to look over the reports at any time. A fair synopsis of every committee meeting should be published for the information of the members."

"Accounts must be audited regularly and full itemized business and financial reports made. Full and accurate information and publicity are essential to true co-operation."

New York   P. H. NORBERG."

A Moreland Report.

During the year 1919 Governor Alfred E. Smith appointed George Gordon Battle, a New York attorney, to investigate the Division of Foods and Markets. He recommended that a milk commission be created with full power to control the milk industry. This would include authority to investigate cost of production, transportation and distribution; and to fix prices with reasonable profits for producers, dealers and consumers. The commission would have five salaried commissioners appointed by the Governor for five-year terms. He recommended that the commission should license all producers' plants, dairy manufacturers and distributors and have power to regulate and control the milk all the way from the cow to the consumer's door.

He criticized the Dairymen's League, Inc., its policies and procedure, and he also criticized Commissioner Eugene H. Porter of the Division of Foods and Markets for accepting assignments and gratuities from the League while occupying a salaried position in the State service. He recommended the removal of Commissioner Porter and also of Charles S. Wilson, Commissioner of the Department of Agriculture. At that time both these commissioners were appointed by the Council of Farms and Markets. The report suggested that Governor Smith request the Council to make the removals. No action was taken on this recommendation.

By that time farmers' experience with political control at Albany discouraged the creation of any new bureaucratic powers. But here again there was a proposal by a distinguished lawyer to strip the dairy farmer of his constitutional and natural right to set the price on the wealth he produces.

Milk Investigated Continually.

During this period the investigation of milk was practically a continuous performance. The plans proposed for the solution of its problems were complex and numerous. At one time five investigations were seeking information.

Governor Smith appointed Martin H. Glynn, former Governor,

THE 1919 MILK STRIKE 157

and Dr. John H. Finley, Commissioner of Education, to investigate the high cost of living. These distinguished investigators gave considerable time to the study of milk and recommended the appointment of a Fair Price Milk Committee to work out a system of economic distribution of milk.

Governor Smith appointed the committee on August 29, 1919. Its members were: Dr. Royal S. Copeland, Health Commissioner of New York City, chairman; Dr. Herman M. Briggs, State Health Commissioner; Jonathan C. Day, former Commissioner of Public Markets of New York City; Francis Martin, District Attorney of Bronx County; Senator Charles E. Russell, of Brooklyn; Lee Kohns, of New York; Preston P. Lynn, of Wanamaker's Store; Miss Sophie Irene Loeb, of the *New York World;* and Mrs. William Randolph Hearst, wife of the publisher.

This committee recommended a State Commission to control and supervise milk from the cow to the consumer. It was to be a State utility. It was to fix the price the dealer would pay the farmer and the price the consumers would pay the dealer. The producers would be required to register with the commission and report to the commission at stated periods, stating the volume of milk sold, to whom sold and where delivered. Dealers, butter makers, cheese makers and other manufacturers of milk products would also be required to register and make reports.

Violation would constitute a misdemeanor and the violation would be subject to a fine of $500 or a year in prison, or both. No action was taken on this recommendation.

The dairy industry has had its own problems but so far it may be thankful that it escaped some of the things proposed for it. But again note that the farmers' inherent rights of ownership of the milk they produce were to be violated.

Borden's Offer Plan.

Not to be outdone by others, the Borden Company proposed a milk plan which was published in the *Dairymen's League News* on January 10, 1920. This plan suggested a special act of the Legislature to confine a corporation to the business of fresh milk distribution in the metropolitan district. It was suggested that the advantages of operation be liberal enough to induce most, if not all, dealers to avail themselves of the profits of it. The corporation would be authorized to issue both common and preferred stock; the preferred stock in payment of net quick current assets of dealers, the common stock for fixed assets.

This corporation would be managed by a board of directors in which dairymen and the public would be represented. Profits would be limited by law and annual financial reports would be filed with the Secretary of State. The proposal, while not carried out in much detail, seemed to visualize a special business corporation limited to milk distributors with State supervision. It would be in the nature of a State monopoly, more or less subject to State regulation. It was a revised Milk Exchange, Ltd. of 1882.

It is proper to note that this plan, like most of the proposals, aimed to strip the farmer of his inherent right to negotiate the price of his milk, and transferred that right to others.

Sheffield's 1918 Profits.

During the Summer of 1920, New York worked itself into a state of hysterics because Loton Horton advised farmers to stop producing milk as he expressed it. He certainly posted advice to farmers to reduce production, but he admitted that in one year Sheffield Farms made a profit of 51 per cent. His statement was:

"I do not believe the fact that Sheffield Farms made $774,000 in one year constitutes a crime. During the year 1918, when we were accused of profiteering because we earned that sum, our critics forgot that the price we paid was fixed by the Federal Food Administration and also the price you (consumers) paid."

Mr. Horton was right about the government price-fixing. In 1917-1918 farmers were urged to produce as a patriotic duty to win the war. The dealers at the time pleaded poverty. It will be recalled that at various times that year the Federal Milk Committee allowed dealers to increase the price to consumers. It made several decreases, and little or no increase to producers. The excuse was to save the dealers from alleged losses. Sheffield Farms cashed in on war prices and its president, Loton Horton, told the truth. There was a lesson for farmers in his blunt candor.

CHAPTER XX

ADMIT FAILURE: SEEK NEW POWER

Fair Promises Broken.

In a meeting of the directors of the Dairymen's League, Inc. on March 22, 1919, Bradley Fuller, attorney, said that the set-up and procedure of the League were fundamentally wrong; that it had not been able to get satisfactory prices due to the fact that there was a surplus of milk at times and the dealers claimed the right to protect themselves against losses incurred or to be incurred; that a method should be provided to care for the surplus, and that a method must be "worked out so that each stockholder will stand on an equal footing in the sale of his milk with every other stockholder."

That was a clear confession that the dealers had been making their own prices, that dealers had resumed their tyranny over producers, and that domination of a farm organization by a ring of self-perpetuating leaders under the name of co-operation was a failure. Producers had known all that for some time. The directors now proposed a plan of organization under Article 13-A of the Membership Corporations Law which was enacted in 1918. In addition to the purposes stated by Mr. Fuller, it proposed to regulate the supply of milk to the needs of the fluid market, to incorporate local associations as the basis of the organization, to protect consumers, and to promote economy and harmony. As given to the press, the plan was to be submitted to producers for their consideration and approval.

This was apparently still following the plan recommended by John Arfman, Milton Lane and myself when we considered what was best to do about the strike; except that the definite provisions of this plan were dictated by the group of officials who had failed in the plan they had insisted on and who proposed to administer the new plan; while our proposal was that the new plan would be worked out by a committee of the ablest producers in the State, and fully discussed and understood before it was adopted, and that the association should be so organized that it could be controlled, not by a self-appointed group, but by farmers themselves. The proposed plan made no provision for the selection of the management, the business policies or the finances. These important considerations, which should be made definite and clear, were vague and uncertain. These defects were pointed out in the agricultural press, and clarification and corrections suggested. In a general way it was what had been approved as the plan of the Department of Foods and Markets at the public meeting in Utica on September 6, 1916 and again at the first annual meeting in December of the same year, and persistently demanded by producers and myself for two and a half years, but persistently flouted and avoided by the League officials during the intervening years of their mistakes, blunders and calamities.

Self-Appointed Leaders.

Instead of waiting for the reaction of farmers, the officials filed their own charter for the incorporation of the Dairymen's League Co-operative Association, Inc. on March 27, just five days after the announcement. It was signed by eight leaders as incorporators.

The directors named in the original certificate of incorporation are as follows:

Roswell D. Cooper, of Little Falls, N. Y.
Frederick H. Thompson, of Holland Patent, N. Y.
Leigh G. Kirkland, of Randolph, N. Y.
Herbert H. Kershaw, of Sherburne, N. Y.
Albert Manning, of Otisville, N. Y.
George M. Tyler, of Honeoye Falls, N. Y.
Frank M. Smith, of Springfield Center, N. Y.
John G. Pembleton, of Owego, N. Y.
Robert W. Siver, of Sidney, N. Y.
Albert L. Brockway, of Syracuse, N. Y.
Earl Laidlaw, of Gouverneur, N. Y.
Bradley Fuller, of Utica, N. Y.

This was a closed corporation of officials and leaders of the Dairymen's League, Inc.

The directors elected the following temporary officers:

President: Bradley Fuller, of Utica, N. Y.
Vice-President: Herbert J. Kershaw, of Sherburne, N. Y.
Secretary: George M. Tyler, of Honeoye Falls, N. Y.
Treasurer: Frank M. Smith, of Springfield Center, N. Y.

We then had the Co-operative Milk Marketing Association and the Country Milk Company, both in the hands of a receiver, the Dairymen's League, Inc., a New Jersey stock corporation and the Dairymen's League Co-operative Association, Inc., a New York membership corporation, all controlled by the same five or six men who had so far persisted in their own blundering autocratic sway. The League had broken down, two of the concerns were in the bankruptcy court, and this last one was yet untried.

Farmers were bewildered by this burden of corporations. They did not understand the phraseology of the new plan. The information given was not sufficient to give them a comprehensive understanding of its scope and effect. To say the least they were skeptical. They asked for information, explanation and discussion. They got nothing.

If dairymen knew then that it was framework for an alliance with Borden's, it would have died "a-borning."

What Farmers Wanted.

Following the 1919 strike from February on, the dealers were clearly in the saddle. To all appearances the management of the League concluded to let matters rest after it admitted failure in March. Farmers were cold to the new plan as proposed. If an attempt to put it in force had been made, the indications were that it would not hold the old membership. The producers grew more and more restless and critical as the management hesitated, and the dealers grew bolder and more domineering. Farmers' demands for reorganization on real co-operative lines were intelligent and persistent. The following excerpts from their letters are typical:

"Some of us did not need the experience of the January strike to convince us that the League needed to be reconstructed. That was apparent to everyone familiar with organization work from the start.

"Why should we have such a colossal plan as this thrown over us like a Mother Hubbard skirt? At the first annual meeting Mr. Dillon called attention to this need, and the members agreed with him and ordered a reorganization, but it was officially defeated. *The Rural New-Yorker* has called attention to this need several times since, and suggested reforms. By all means let us have it now. There is nothing new in these organizations except to adjust details to peculiar needs. Co-operation has been in successful operation for years in other places. When the members retained control, they succeeded; when the control became autocratic, they failed. Why lose time and money to repeat the mistakes others have made? Make the organization co-operative. Bring in your best men with successful business experience. Let them devise a plan and then give every member an opportunity to vote his approval or objection."

New York   E. M.

"I am interested in the Dairymen's League. I am, however, opposed to the plan for organizing the Dairymen's League Co-operative Association. The only reason why milk dealers maintain receiving stations and manufacturing plants is for the purpose of getting milk to distribute in the large cities, which is the profitable end of the business. Under the proposed plan producers who purchase and operate the receiving stations and manufacturing plants will still leave the dealer in control of prices and the profitable end of the business."

New York   C. W. HALLIDAY.

"I am glad to see a farm paper that has the common sense to wish to discuss such a gigantic project as the League has undertaken before advocating any one plan, and I heartily commend you for the policy you have taken.

"The League has never asked its members to submit any plans or ideas on any subject, and the leaders of the League should give heed to *The Rural New-Yorker*'s suggestion and example in that respect."

New York   GEORGE C. PORTER.

"The present management of the Dairymen's League is too expensive and does not get results. There are too many officers drawing big salaries and too many conferences with dealers. Let the dealers alone. Do not chase after them. The proposed plan has not the confidence of the

ADMIT FAILURE: SEEK NEW POWER 163

dairymen. It is too complicated. It takes authority away from dairymen and places it with a very few, who are given too much power. Let the dairymen own and manage their own plants, equipped to make butter, cheese or pasteurized milk to ship. Dairymen are well able to manage their own plants."

New York H. T. FIELD.

"After all that *The Rural New-Yorker* has done and is doing for agricultural interests, we farmers should hardly need the incentive of a prize offer to express our opinions. Dairymen have built and equipped local plants. The new movement being agitated seems to be to do away with plants just recently acquired and give up what we have gained of independence, and leave us again to trust to the honesty of a set of middlemen. Let us keep our own local plants and equipment."

New York ANDREW J. SPERLING.

"Any radical change affecting dairy organization should be voted upon by dairymen. The directors will then be properly instructed. Local branches should be under local farm control. Business accounts of the League should be audited and the books and other records open to the members."

New York W. J. DURKEE.

"I cannot see anything fair in the milk situation. I held my milk in the strike. I had been getting $5.00 a day from a condensery. Now Borden's rejects my milk and the milk of between 20 and 30 farmers around here who are in the same predicament. Why is not every League member protected?"

New York HENRY E. LINK.

"Two organizations as proposed would defeat farm co-operation. Complete unity must be the keynote of dairy organization. Two organizations would destroy this unity. Real co-operation is essential. No successful local co-operative plan should be in any way disturbed."

New York EDITH ANDERSON.

The Annual Meeting.

The annual meeting of the Dairymen's League, Inc. for 1919 was held in Jersey City, New Jersey, on December 20th. President Cooper said prices for milk did not cover cost of production, but expected that a new plan of computing would increase prices about thirteen cents a cwt. over the old method. The new method was based on average prices for butter and cheese with seasonal variations long used by the dealers. Mr. Cooper said the present plans of the League "could not be permanently successful" and he thought "the co-operative plan proposed in March or something like it would be necessary."

John D. Miller, in his capacity as full-fledged boss of the organization, pleaded for the retention of the executive officers. It was easy legally to control the organization by proxy votes of the stockholders, but the management had been freely criticized and R. D. Cooper's position had not been very secure from the first. Many suggestions for a sweeping change in the management had been made during the year. Milo Campbell of Michigan was again on hand to boost President Cooper as he had been since and including 1917. He again made an appeal for the executive officers. No organized opposition appeared and for the most part the official slate went through.

A charter for a new organization had been executed and filed by the official group on March 27th. A co-operative plan had been proposed by the official group the same month. This would seem to have been the opportune time to explain it in full detail and to have discussed it in the open. No time was given to organization policies or principles, which was the need rather than emotional propaganda. The general sentiment, however, reflected something of the feeling of dairymen throughout the milk shed.

As usual, resolutions sent in and proposed by members and delegates were required to be in writing and referred without reading to a resolutions committee. Comparatively few of them ever got back to the convention. The committee summarized those the management wanted or was willing to accept. The intent to keep the new organization under the old management was indicated in a resolution to the effect that the Board of Directors of the old and the new organization should be the same.

That purpose was also indicated in a statement by Mr. Cooper that the directors at a previous meeting had authorized that

Dairymen's League, Inc. money be used to organize the Dairymen's League Co-operative Association, Inc.

It was also provided that Bradley Fuller be added to the legal staff at a salary of $8,000 a year. However, a resolution to increase the revenue by collecting two cents per one hundred pounds from producers instead of one cent as formerly, was voted down. A resolution to limit payments to directors for attending meetings reflected strong producer sentiment for economy, but it was finally laid on the table.

The treasurer reported:

Income for the Year 1919	$294,274.44
Total Expenses	344,499.60

Deficit for the Year		$50,225.16
Big items in the expense were:		
Official salaries	$ 34,964.97	
Official salaries	23,231.43	
Legal (deficit)	26,483.17	
Dairymen's League News	46,916.63	
Stockholders' meetings	21,022.12	
Advertising	21,675.74	
Mass meetings	12,784.64	
Directors' Expense	14,000.50	
Directors per diem	5,828.12	
Organization	16,858,17	
Field	34,733.07	

CHAPTER XXI

THE MAJOR MILK TRAGEDY

Farmers Distrustful.

It just seemed as if everyone was in a conspiracy to ruin milk producers during the summer and fall of 1920.

During the late months of 1919 the milk supply fell off and the demand increased prices during the period of scant supply, which extended into 1920. But prices declined from $3.69 in January, 1920, to $2.55 for April, or nearly two and a half cents a quart. The dealers' spread was $4.50 a cwt. At that time Grade B milk retailed at fifteen cents in quart bottles and twenty cents a quart in pint bottles. Grade A was eighteen cents and twenty-two cents in pint bottles. That gave the farmer five and four-tenths cents a quart to the dealer's fifteen cents, a spread of nine and six-tenths cents.

Dr. Charles E. North reported, after long investigation under the auspices of a Rochester committee seeking to reduce the cost to consumers, that farmers supplying the city of Rochester with milk had lost one and eight-tenths cents a quart for a previous year. Farmers, he said, worked for eleven cents an hour. Feed had gone up to $70 a ton, and it was generally admitted that farmers were not receiving cost of production.

I published a typical farm letter at the time. It serves to show farm sentiment and judgment, as follows:

"We seem to have no definite and generally accepted milk dairy policy. About a year ago Bradley Fuller, speaking for the organization, said that the policy of the Dairymen's League was fundamentally wrong. It has not been changed since. The Dairymen's League Co-operative Association, Inc., then organized, has made little or no progress in more than a year, and has not been generally approved. A vote would no doubt show a large majority of farmers opposed to it. With such opposition or uncertainty, to say the least, it cannot be successful and must ultimately cause confusion if not conflict of interests. The dealers have the long end of the stick at the present time, and we need a policy on which all producers can unite to defeat them. *The Rural New-Yorker* has given long and careful study to this problem. The original plan was successful in the first fight with the dealers. My recollection is that the plan you published two years ago was full and comprehensive."

E.N.
New York

The Pooling Contract.

The pooling contract seems to have been developed behind closed doors during the early part of 1920. There was no reference to it when the failure of the old League was announced in March, 1919. Little was heard from the proponents of the new plan and nothing about the contract until the Spring of 1920. The contract was not then published and copies of it were not generally distributed.

The management sent canvassers out in the dairy counties to solicit farmers to sign the contracts. This brought inquiries. Farmers asked me what protection they had if they signed it. To this inquiry I answered that it all depended on whether dairy farmers themselves had control of the corporation. If so, the details of the contract would not be important because any hardships in it would afflict all producers, and the members could and would change the provisions of the contract to suit themselves. If, however, the charter and by-laws were not definite enough to give the members control, the contract could not be signed with safety. The officers might usurp the powers of the members as had happened in the former organization. Of course, I took it for granted at the time that all producers were to be members just as the farmers did themselves. I recall no definite statement asserting that farmers would constitute the membership, but the general expressions indicated that such was the intention, and no one even thought of any other possibility.

I stressed farm control in every instance, but always advised co-operation with full information for each and every producer, and a ballot for each member on important questions, such as election of officers, general policies of the organizations, the rais-

Conditions Could Not Be Worse.

On September 28, 1920 the League officials issued a statement saying that they had accepted a price forty-one cents a cwt. below the cost of production. Furthermore, the dealers gave notice that they would not keep all their plants open nor take all the milk produced by their patrons. The essence of the long statement was that the dealers dominated the situation. Compared with butter and cheese prices milk had never before been sold so cheap. It was estimated that 25 to 30 per cent of the milk in the State had no market.

There was what we now call a "recession" that year, but that does not account for the wide spread in milk prices between producers and consumers with no attempt on the part of the organization to protect the producers. The relations between the League officials as spokesmen for distressed farmers and the Borden officials as leaders of the dealers' organization were most cordial.

The League opened a retail distributing plant in Newark, N. J., that year. C. A. Weiant, a Borden man, got the job as manager. There may not have been any Borden interests in this undertaking; but if the purpose was to show the weakness of the then existing organization and to make farmers desperate, this affiliation might serve their purpose. Whether the joint interests were intentional or a mere result of incompetence, the effect of it was to induce some farmers to sign the proposed "pooling" contract, on the theory that it could not make conditions any worse.

Extravagance Begins.

In the early part of November, 1920, a statement was given out by the League officials intimating that a drive was under way for signers of the "pooling" contract, that 20,746 dairymen had signed it, and that pooling would begin when 50,000 members had signed the contract.

The "pooling contract" had not been published for the information of farmers nor publicly discussed. Copies of it were not readily available. When farmers signed it they got no copy. An occasional farmer refused to sign it without a copy and canvassers promised that a copy would be sent direct from the office. This was not done. Later when a farmer withdrew and demanded the return of his contract, the request was refused. The drive was conducted by canvassers who had been trained for the work and sent out to the dairy counties to persuade farmers to sign the agreement.

Conditions in the producing sections were discouraging. Prices had dropped far below the cost of production and many farmers had no outlet for their milk. The management had completely broken down in its salesmanship and was devoting itself to the promotion of the new organization. The expenses reported at the 1919 annual meeting showed that the official salaries for the year were close to $60,000, the legal fees in excess of $26,000 and the organization and canvassing in excess of $50,000. The *League News*, which was published for propaganda work, had a deficit of close to $47,000 in addition to the 50 cents subscription charged up to every producer. For the organization as a whole the farmers paid that year $344,499.60. Farmers did not understand the set-up of the new association, but they did not like it. They did not know what the pooling contract was all about, and nobody knew what it meant well enough to explain it to them. Lawyers shrugged their shoulders, and from a legal standpoint they condemned it, yet because of the existing conditions few definitely advised their clients to stay out of it.

The 1920 Meeting.

It is doubtful if any lay dairyman knows today whether the Utica, New York, meeting in December, 1920 was a meeting of the members of the Dairymen's League, Inc., which according to its charter had to be held in New Jersey, or a meeting of the Dairymen's League Co-operative Association, Inc. It didn't matter. The twenty-four exclusive members and identical directors had their paid agents canvass the whole State, and had gathered in enough of their followers to control the meeting.

To all appearances, it was a representative dairy meeting. It included probably a majority of cool-headed, usually deliberate acting, sober-minded farmers who were in favor of a co-operative and pooling organization, but opposed to the contract. The milk trust and many dealers had become emboldened by the helplessness of the League for two full years and had created such an intolerable situation that many farmers would accept any change as better than the then prevailing conditions.

The Nestle Company had been especially odious. This situation was dwelt upon by the speakers. John D. Miller, now the recognized motive power behind the League officials, made the most of the farmers' resentment to the dealers' tyranny. He made the definite statement that the "open conflict is on between middlemen and farmers." Coming from him at this time, in connection with his assurances that it was to be, as he put it on many occasions speaking to farmers, "your organization," it appeared that the management had at last assumed a real farm complex, that it was at last ready to conduct a real co-operative farm association for the promotion of an economic distribution of milk, and that farmers would be in a position to direct the policy and procedure of their organization.

Pooling Contract Adopted.

The active element, however, was strictly emotional. All producers had felt the sting of the dealers' lash. They recalled what had been done four years before. The leaders didn't want to discuss forms of organization, policies or procedure. They wanted to fight, to spend money and go ahead. They didn't want to hear anything about accredited principles of co-operative organization. They didn't care whether they were organized from the top down or the bottom up. They didn't care about established form or perfection of detail. They didn't count the cost. They were pepped up and they were now demanding action. They were acting. The spirit was infectious. The officers offered the floor to anyone who opposed the proposed set-up or the contract. Those opposed did not speak. It would have been useless. The resolution was a clear-cut definite approval of the pooling of milk and of the contract as it stood without change, reservation or limitation of any kind. It was approved without opposition.

Yet, a larger majority of the audience were clear-headed, conservative dairy farmers. They preferred more deliberation and a better understanding of what it all meant. They didn't understand the set-up or visualize how it was to work. They had misgivings about the contract. They were opposed to it. But none of them knew that the corporation was limited to twenty-four members. The worst they feared was a repetition of the failures of the past. They had no suspicion of bad faith.

They held to the original 1916 creed:

"We are willing to work with them [the milk dealers] hand in hand, and we will be glad to see them fully rewarded for the time and capital necessary to distribute milk, but they must no longer tyrannize over us. They must not exact unreasonable prices from the consumer and thereby cut off our outlet for our product. There must be no more "2 cent-milk" for the producer and at the same time "12 cent-milk" to the consumer. They must not drive independent dealers out of the distributing of milk and they must not keep new men out of the business."

I was, of course, in favor of a co-operative organization. As far as I know, I was the first to suggest pooling the milk and make provision in the co-operative plants to manufacture milk not required for the fluid market, and for milk not inspected for shipment to the city. But I had been insisting on protection for producers in the charter, and provisions and modifications in the contract to safeguard the interests of dairymen. Several of my friends among the directors and laymen appealed to me to support the new organization. I asked them the straight question: "Are farmers to have a monthly profit and loss statement accounting in detail for the milk and the expenses with their returns?" They assured me that such an accounting would be made. With this provision I felt that farmers could check up on the business as it affected them, and that while I could not advise farmers to sign such instruments, I felt that New York dairymen with the evidence before them would not long stand for extravagance, dictation or exploitation.

The new board elected George W. Slocum, of Milton, Pa., as

president to replace Roswell D. Cooper. Our 135,000 New York dairymen then had one New Jersey corporation and three New York corporations, of which two were in the bankruptcy courts, and all of them under one identical board of directors and officers, the president, vice-president and general counsel being citizens of Pennsylvania. The members of the new association were twenty-four in number, and self-appointed. They elected themselves as twenty-four directors and then elected their own officers, but in talks to farmers it was always "your organization."

Even as a citizen of another State, the election of Mr. Slocum afforded opportunity of hope. I interviewed the new president. He told me that "We must have better co-operation from the head office right through to the membership than before." He would organize local organizations and furnish prompt and full information of just what was being done in the meetings of the executive committee and of the board of directors. He promised a regular synopsis for the press.

"The new administration," he said, "is not to be a one-man affair; but a real functioning of the whole organization. The foundation of the work would be the pooling plan and an inexpensive central plant to take care of any surplus milk when it appears."

In principle and plans he seemed to prefer to start at the top with a centralized control with a spread of power and authority down to the producers, while true co-operation, as I have always visualized it, begins with the individual producers on the farms as a foundation, organization of local associations and an affiliation of them to form a central body receiving its authority from the producers. But he seemed liberal even with this policy and expressed himself as willing to listen with an open mind. He invited suggestions, discussions and criticisms. He impressed me as being the type of farmer needed for co-operative work. I said so in an editorial.

Pooling Contract Described.

Some call the pooling contract ironclad. Others tell that the like of it could not be found in the records of jurisprudence.

THE MAJOR MILK TRAGEDY

Thousands of farmers flatly refused to sign it. Some of them were persuaded to do so.

Willard R. Pratt, Utica attorney, in a brief in the Holmes case, which I discuss later, briefly stated some of the provisions of the contract as follows:

1. Producers agreed to deliver all their milk to the League to dispose of as it pleased.
2. The proceeds of all the milk received by the League are to be blended into a common fund.
3. If the League fails to sell the milk, the producer will himself manufacture same and deliver it to the League or pay over to the League all he receives therefor.
4. All offers made to the producer for his milk must be turned over to the League.
5. Out of the common fund the League may deduct such sums as it charges others for like services and which it deems necessary to deduct to pay expenses already incurred, interest on indebtedness, the League's overhead charges, depreciation, guarantees made by the League and all other expenses the League shall estimate as essential to carrying on the business.
6. As to such deductions no certificate shall be issued to the producers and hence no part of the same shall ever be paid back to the producers.
7. In addition to the above, the League may make further deductions to create a fund to retire and pay up loans made by the League in building warehouses and other buildings; to purchase equipment and furnish such working capital as the League desires; such deductions to continue as long as the milk is delivered to the League, and at the end of each fiscal year the producers shall receive a certificate of the amount so deducted in such form and payable at such time and at such rate of interest as the League shall determine.
8. The League shall make distribution subject to such differentials between producers as the League shall determine and that the League shall be sole arbitrator of such distributions and the same shall be final and conclusive on all parties.
9. Producers in the absence of fraud waive all their right to an accounting by the League, both at law and in equity and the contract shall be a bar to any proceeding for the same.
10. The League may borrow from time to time such sums of money as it desires and to pledge for the payment thereof any unsold milk, dairy products, accounts receivable, notes, and trade accep-

tances in which the producer has any interest as owner or otherwise from the proceeds of which such borrowed money shall be paid.
11. Each producer must subscribe for the *Dairymen's League News*, the subscription price to be deducted from his milk bills.
12. The contract shall run continuously unless notice to the contrary is given in writing by the producer between February 14th and February 28th.
13. The producer covenants and agrees to and with the Association that if he at any time refuses or neglects to deliver such milk or the manufactured product thereof produced or manufactured by him to the Association, or upon order, at such time and place as the Association may direct, then and in that event in every such case the producer neglecting or refusing so to do will pay to the Association for such refusal or default, the sum of ten dollars per cow for thirty cows, and if such default or refusal shall continue for more than one month, an additional sum of three dollars for each cow per month, for the same number of cows, so long as such default or refusal continues, none of which payments are to be construed to be a penalty or forfeiture, but as stipulated liquidating damages as prescribed by Section 209-A of Chapter 655 of the Laws of 1918 of the State of New York, and it is hereby agreed that the Association will suffer by reason of such refusal or default.

A One-Sided Contract.

Besides all this, payments for every month's deliveries were not to be made until the 25th of the following month. This meant that the farmers furnished a major portion, if not all, of the capital to finance both the wholesale and retail distribution of the milk under this plan.

If the Borden contract with producers, which had been in use for some fifty years, was not used as a model of the pooling contract, the authors of both certainly had similar purposes in their minds. That purpose in both cases was to bind the farmer to do everything for the comfort and benefit of the party receiving the milk and to bind the other signers to the instrument practically to nothing. In the Borden contract the farmer bound himself to do everything that Borden's lawyer could define on two pages of legal paper. The only thing that Borden's bound themselves to do was to pay for what milk they received and used at the price that they fixed themselves.

In the pooling contract the farmer bound himself in fifteen different obligations and the League bound itself to absolutely nothing except that it would make a return the 25th of the following month for the milk received during the previous month, but assumed no obligation for any price and reserved to itself the privilege of using any or all of the returns for promotional or capital purposes. The only other obligation the League assumed was the mere formality of giving the producer after the close of the fiscal year a certificate for the amount of money deducted for capital purposes. This was to be used as a revolving fund for a period of five to ten years which the officers of the League would fix and at a rate of interest which the officers would determine.

As a matter of fact, the League was organized to do business without profit, and this loan for which the certificate was given could not be repaid until the money was deducted in sufficient amount from the monthly returns, so that the farmer was paying his own loan and his own interest by monthly installments. He was obliged to wait until the end of the year for the return of his contributions for interest and obliged to wait until the end of the period, which has varied from five to ten years, for the original loan. In the meantime his deductions continued monthly and indefinitely and his loans increased to the end of the period. The loans totaled approximately seventeen million dollars.

Farmers Were Not Members.

The court in the Holmes case found no mutual provision in the "pooling" contract with the possible exception of the statement that the producer wished to have the service of the corporation. So phrased it could be interpreted as an expected benefit to the producer, but actually the farmer's patronage was essential to the corporation, a benefit to it and its 24 members, but the phrase saved the contract from being declared unlawful and void, which in truth it is believed by legal authorities that it then was and probably still is. Farmers were not members. Their relations to the 24 leaders' corporation were carefully stipulated in the contract. They agreed in the contracts to many specific obligations favorable to the corporation. There was nothing in it to safeguard the rights of milk producers.

It also seems clear that the leaders had determined not to make a full accounting of their trust as an agency for the sale of milk. This would seem to be true from the following provision of the contract:

"The producer, in the absence of fraud, hereby waives all his rights in law or in equity to an accounting therefor, and this contract shall act as a bar thereto in any proceedings taken by the Producer therefor."

The lawyers who wrote that contract well knew that a farmer could not prove "fraud" when he barred himself from an accounting.

The law requires co-operatives to file an annual report with the Commissioner of Agriculture. For a time incomplete annual reports were filed and published. But public analysis of the reports revealed that items were lacking in published reports, making them practically worthless. In answer to this criticism, the leaders stopped publishing the annual reports and had a provision inserted in the law restraining the Commissioner of Agriculture from allowing access to the annual reports without permission by an officer of the corporation.

Farmers Sold Worthless Stock.

The officers decided to abandon the Dairymen's League, Inc. in March, 1919, when they admitted failure. At the same time they had filed the charter for the new Dairymen's League Co-operative Association, a membership, non-stock corporation. But they continued to canvass for sales of the stock of the Dairymen's League, Inc., and collected money on such sales as late as June 21, 1921. Of course, it was necessary directly or indirectly to mislead the farmer to sell him stock at that time. The stock did not have even a speculative chance. Evidently it was void of value or service. As late as 1938, a New York farmer who bought four shares on June 21, 1921, supposed it had some value. I thought George W. Slocum, who signed it as president, might feel that himself or his associates might be willing to refund the cost, but they declined the privilege of so doing. Of course, the stockholder has no legal redress now; but I believe few farmers will consider it an ethical, moral or honest transaction.

The twenty-four directors held on to the offices of the Dairymen's League, Inc. for several years to make it impossible for the old remaining members to operate it for their own purposes. No meetings were held and finally the State of New Jersey annulled its charter because of the failure of its officers to pay the corporate tax.

CHAPTER XXII

POOLING AND CLASSIFICATION BEGIN

Classes and Blended Prices.

On May 1, 1921, the Dairymen's League Co-operative Association began to pool the milk of dairymen who had signed the pooling contract. The beginning of pooling was also the beginning of the classified plan and blended prices.

A strenuous campaign had been put on to induce unwilling farmers to sign the pooling contract. Most farmers resisted so the board of twenty-four members kept themselves in control of both organizations and attempted to sell milk for the two dairy organizations. For the Dairymen's League, Inc. they fixed a flat price for May, 1921 of $2.30 a cwt. for milk of 3 per cent fat test.

For the Dairymen's League Co-operative Association, Inc. they fixed the classified prices for the pool patron as follows: Class 1 $1.95 per cwt.; Class 2 $1.65 per cwt.; and Class 3 $1.50 per cwt. Class 1 was fluid milk, Class 2 was milk to make cream, and Class 3 was milk to make butter and cheese.

The blended price returned on June 25th was $1.41 per cwt. or 89 cents less than the flat price paid the old League members.

The retail price in New York City was: Loose milk in stores, per quart ten cents; Grade A in bottles eighteen cents; Grade B fifteen cents; and Certified milk twenty-eight cents.

This gave Borden's for distribution a spread of 12 cents a quart or $5.64 a cwt. compared with 3 cents a quart or $1.41 to the farmer for production. These results revealed Borden's influence and purpose.

To quiet the complaints, instead of increasing the low price, the officials sent a letter in September 1921, to all buyers of milk produced by members of the old Dairymen's League, Inc. giving the dealers the option not to pay their producers the price pre-

POOLING AND CLASSIFICATION BEGIN 179

viously fixed but to pay them the lower blended pool price, the dealers to retain the difference. The difference for the month of October, 1921, was eighty-six cents.

At that time these non-pool farmers were contributing $8,000 a month for the services of these directors who voluntarily authorized their dealers to cut the October price to be paid these farmers eighty-six cents a hundred pounds. These farmers said they believed that the cut was made to punish them for refusal to sign the pooling contract and to force them to do so. It was understood at the time that to the everlasting credit of the dealers buying the non-pool milk they refused to exercise the privilege and paid their producers the full price as originally agreed.

The classified plan was announced as something new and a great discovery for dairy farmers. As a matter of fact it was virtually the plan or formula used by the Borden Company for years to estimate prices to be paid farmers for six months periods. With the aid of previous records the Borden Company could estimate their requirements of fluid milk and cream in advance for the six months. Borden's could also estimate the probable supply. They knew what the price of butter and cheese would be because, according to Federal Trade Commission reports, they are one of the four concerns that make the cheese prices and one of the six that fix the butter price. The only thing they had to guess on was the weather. They overcame that difficulty by fixing the monthly price by the formula as low as their conscience would admit and then taking off some more by way of self-insurance.

There were, however, some disadvantages to dealers in fixing the price in advance. If the price was too low, farmers had a chance to reduce production; an unexpected drouth might automatically reduce production, and in either case it might be necessary to increase the price to get their requirements of milk. Again, consumption might fall off or increase for one reason or another and there would be no way to charge the farmer for the loss of profits. His price was fixed and he had a choice of keeping up the supply or reducing it. If the price set from month to month were too low, farmers would go into a rage and perhaps strike. With the classification scheme the farmer would not know what the price would be until he got his return. The fluid price would look good and hope for the best would keep the farmer going from month to month at top speed of production. Besides, when farmers complain of low prices it is convenient to have the surplus argument so the dealer does not need to worry about the "returns" from the wagon routes. Everything that is not sold for fluid use goes into low price classes.

The Federal Trade Commission found some of our biggest dealers in Connecticut and Philadelphia chiselling fluid milk to consumers and returning a low class price for it to the producers. Yes, the classified plan is a find for the distributors!

Under the plan the farmer ships his milk to a dealer on consignment, waits an average of forty days for his money, authorizes the dealer to fix his own price for it and waives his right to an accounting for milk or money.

During the summer of 1921, it was rumored that the Dairymen's League, Inc. was about to be abandoned, so producers who did not sign the pooling contract would not be able to sell milk to the plants they had helped build and maintain. Some three hundred producers held a meeting in Oneida County, and appointed a committee to protect their interests. They drew up and published many strong suggestions for continuing their organization on the co-operative plan originally proposed in 1916.

Dairy Farmers Protest.

As a general rule the official program went through the League meetings, local or State, without opposition or debate. At the annual meetings a delegate who rose to speak would be politely advised that he would be heard at the proper time, but just then he was out of order. Sometimes he would be instructed to put his message in the form of a resolution. The "proper time" never came and the resolution was not reported. At local meetings generally, the member who made a complaint or a suggestion was appeased with fraternal soft talk. If he persisted, he was embarrassed in one way or another.

But the Oneida County meeting on November 19, 1921, was an

exception. Many old League members refused to sign the contract because it contained odious provisions. Others had signed the contract believing that it carried membership in the new co-operative organization. They had later learned that this was not true. Again, in October, 1921, the identical directors who controlled both organizations had fixed a price for pooled milk to supply the local Utica market at $3.37 per hundredweight. At the same time they fixed the price for non-poolers at $2.27 for the same inland market. This was a discrimination of $1.10 against the farmers who refused to sign the, to them, odious pooling contract.

A resolution was presented criticising the conduct of the management, for discrimination against old League members, directing their delegate to present the resolution and the facts to the annual meeting, and requesting Governor Nathan Miller to investigate the pooling association and the dispositions made of League money by the directors. Bradley Fuller and E. R. Eastman, then editor of the *League News*, refused to explain the reason for the discrimination in price. They also refused to explain why the farmers who signed the contract were not members in the Association, or why the membership was limited to the 24 directors. Mr. Fuller was asked in open meeting why the directors had opposed, in the Legislature, the Everett bill which required co-operatives to make an accounting to members. He refused to answer these questions.

The pooling members voted against the resolution but it passed by a big majority vote. The non-poolers also elected all the officers and a delegate to the annual meeting which was held in New Jersey the following month.

Cooper Sent Back Home.

About this time R. D. Cooper, having been relieved of his duties as president of the Dairymen's League, Inc., went to Wisconsin to set up a dairy organization there on the pattern of the Dairymen's League Co-operative Association, Inc. I never knew who inspired this undertaking on his part, but I suspected at the time that it was suggested by his former mentor to get him out of the way. At the same time, if Wisconsin dairymen could be induced to accept a centralized autocracy to control their business, the example of Wisconsin would be an argument to use on New York dairymen who opposed the "iron-clad" contract. But Mr. Cooper was not successful in his mission.

In Wisconsin a policy was approved for a state-wide co-operative system. It began by organizing farmers in local associations and affiliating them to form their state body. In this plan farmers owned their plants and controlled the local units, and through the units they also controlled the state association. An attempt, however, was being made to change this policy by inducing Wisconsin dairymen and other farmers to create one centralized corporation after the pattern of the new Dairymen's League Co-operative Association, Inc. of New York. It would own all plants and equipment and do all the business, leaving farmers at home to work and giving the leaders a free hand with the money.

The Wisconsin plan was already in operation elsewhere. It is the plan that succeeded in Denmark and with citrus fruit growers in California. It was succeeding in Wisconsin.

When it was discovered that Mr. Cooper had become a promoter in Wisconsin, Prof. Macklin of the Wisconsin Agricultural College objected to the New York plan as "fundamentally unsound, undemocratic and impractical in execution." In effect he called it a plan to benefit promoters who had no concern but their own interests. It opened the way to waste and extravagance, he said, and denied local men the opportunity to develop by actually doing their own work. In short, Mr. Cooper was told by resolutions of practically all the farm organizations of the State that there was no place for him or his despotic policies in Wisconsin. There were no brass bands to greet Mr. Cooper on his return to New York the next day.

Farmers Not Association Members.

During the month of November, 1921, I received in the mail a communication which contained the following statement:

"The producers who signed the pooling contract have been led to believe that they are members of the co-operative association. This is not true. They are not members. They have no voice or control over the pooling association. ****** The relation of dairymen is simply that of individuals entering a contract with a corporation. The association consists of only a handful of men, most of whom are appointed officers, who have the right under the contract to make the price of milk, to collect for it and take out whatever amount they please. They own and control millions of dollars worth of plants and equipment purchased by the producers' money. The dairymen are required to sign away their right to an accounting."

I went direct to the League office. I found President Slocum and Bradley Fuller together. I asked them exactly how many members there were in the new League corporation. Both of them hesitated to say whether it was twenty-four or thirty-four. They agreed that originally there were only twenty-four but seemed not to be certain whether former directors were yet members or not. They admitted that the farmers who signed the pooling contract were not members of the association. They told me that the fiscal year ended March 31st and the annual meeting (of the twenty-four members) was held in June. "It is a closed meeting," they said, "and a public meeting has not been considered." Both men were sure that they never said anything to lead anyone to think that dairymen who signed the pooling contract were members. The association owned all the plants and equipment and the business was all done by it, including the selling of the non-pool milk. The old League functions were negligible, they said.

The worst was admitted. I had not believed the statement. I had felt there must be some explanation. There was not. It was all too true. In my worst fears I had never visioned such a stupid, selfish intrigue. I knew some of the 24 men. There were honest men among them, but these were inexperienced. The executive committee dominated the board and John D. Miller dominated the committee.

I knew that he and the committee had adopted a rule-or-ruin policy from the first. If a policy was good for milk producers but not so beneficial to themselves it was not adopted. I saw them sacrifice the producers' interests to increase their own importance and powers. But I had thought such procedure due to ambition, selfishness and the conceit that led them to believe that anything they favored would be best for everybody. But here was intentional usurpation admitted, confidence was betrayed and a trust violated. No hope could be held for the dairy industry with such leaders, and yet they would brazen it out and wreck the whole structure rather than give up the power they had usurped. I had an emotion akin to pain. Our great dairy industry was "on the block."

But I knew that New York dairymen had successfully organized once. They had had a taste of co-operation in its simplest form. They had relieved themselves for a time of the tyranny of a group of practical middlemen and I believed they would restore their organization to the form, procedure and services originally intended. That meant the development of their own latent talent and ability in the direction and control of it for themselves. So far they had been defeated in this purpose by the destruction of their unity through misleading propaganda, by an abuse of their credit and confidence and above all by the surrender of the fruits of organization to the Borden Company. But my faith in their "horse sense" and my hope in ultimate justice are yet firm. Unless the spirit and character of New York dairymen are destroyed they will, I believe, again recover control of their own business and operate it with success.

Leaders Destroy Unity.

The annual 1921 meeting of Dairymen's League, Inc. was held in Jersey City, New Jersey, in December. President Slocum was unable to be present because of the death of his wife. John D. Miller, vice-president and counsel, presided. The official group had full and exclusive control of the organization machinery. Its program was laid and strictly followed to the end.

A group of members who refused to sign the contract presented a prepared resolution which was referred to the resolutions committee. In its preamble it stated that there were then three organizations of dairymen in the State, viz., Dairymen's League, Inc., Dairymen's League Co-operative Association, Inc., and the Non-pooling Co-operative Association, Inc. It laid down the principle that one united association was essential to success. The resolution therefore proposed that a committee of fifteen producers be appointed to work out a plan of reorganization to which all could agree. If a unanimous agreement could not be reached within sixty days, then the points on which they disagreed would be clearly expressed and submitted to all organization members in the milkshed in the form of a referendum ballot, conveniently arranged for a vote in person or by mail; the referendum to be conducted and the votes canvassed by the committee. The decision of the majority was to be accepted by all.

After considerable dispute a few minutes were allowed the protesting members to state their position. W. R. Pratt of Utica spoke briefly for the resolution. H. B. Sweet, the delegate from the Oneida County local, complained that non-poolers had been discriminated against in the sale of their milk by the directors to the amount of nearly a dollar a hundred pounds and further denied the privilege of marketing their milk through the plants they built and paid for and were supposed to own.

The resolution was not reported out by the resolution committee. Hence, it was not voted on. As a practical matter the officials had the lists of stockholders, and their money to reach them by person and by mail. They had proxies and followers enough to approve or disapprove anything as they wished.

The principal item in the day's program was a resolution authorizing the directors of the Dairymen's League, Inc. to notify its members that the corporation would not sell their milk after April 1, 1922. It went through without a protest—except for the small group who had come to the meeting with their resolution to appoint a committee of fifteen. A major part of the members were neither there nor represented. They were disenfranchised. Their ship was scuttled by the officers and crew. The once completely united New York dairymen were split wide open by the men whose duty it was to keep them united.

CHAPTER XXIII

THE BORDEN-LEAGUE ALLIANCE

Borden's Press the Button.

On April 1, 1922, the Dairymen's League, Inc. stopped selling milk for its members. Its twenty-four directors were also the directors of the Dairymen's League Co-operative Association, Inc., which for a year had been selling milk for dairymen who had signed the "pooling" contract.

The Borden Company, which operated a large line of plants, announced that it would buy all of its supply for the New York metropolitan market from the Dairymen's League Co-operative Association, Inc. Borden's had a large number of producers who objected to the contract. Some of them had patronized the Borden plants for twenty years or more. Many of them had originally contributed to the building of the plants for the sake of a local market. But Borden's were determined to put the new corporation in business. Farmers were told to sign the "pooling" contract or keep their milk at home.

The time was short. The date was chosen with a purpose. May 1st is no time for farmers to be without a place to ship milk. They signed. So Borden's stunt put the Dairymen's League Co-operative Association, Inc. on its feet as a going concern. A small number of other dealers contracted with the League for their supply of milk, but the major number wary of Borden's domination, refused to deal with the League and continued to buy from their local producers who balked at the "pooling" contract.

Among these dealers were Sheffield Farms Company, Empire Dairy, Model Dairy, Evans Dairy, Keystone Dairy Company, Newark Milk and Cream Company, M. H. Renken Dairy Company, Willowbrook Dairy Company, and others totalling 200 or more.

Ruled With Autocratic Power.

No despot ever ruled with more autocratic power than the group of about five men who dominated both the old League and the new organization through one board of twenty-four directors. They assumed their power in the old League by virtue of a few proxies easily obtainable through their official connections, and constituting the whole membership of the Dairymen's League Co-operative Association, Inc., no dairyman had any authority to check their sway in any way. Nothing but a spontaneous revolution of all the producers could stop them. When they began the organization of the co-operative the secretary's report showed that the old League had a surplus of $175,000. This was an accumulation of annual surpluses in excess of the expenses from the sale of stock and the revenue of one cent per hundred pounds of milk contributed by dairymen. In addition to the revenue thus collected, the officials had continued to sell stock in the old League, long after they had announced that it had failed and after the co-operative charter was filed in 1919 to replace the old corporation. A little thing like morals or ethics did not seem to bother them.

The officers held on to the control of the old League after it had ceased to function and the members complained that they never accounted for the large sum of money last reported in the treasury. The loss of the money was not their only misfortune. The experience discouraged them, and gave them a feeling of distrust. It discouraged them from further organization among themselves.

"What is the good," they would say. "We had an organization once. It was successful for a time. The money we had in it was stolen from us and what assurance have we that if we organize again some one will not do the same thing to us. We would rather be overdone by strangers or by enemies, if you call them so, than by those who pose as our friends."

Why Farm Unity Was Destroyed.

No one would accuse the Borden Company of making this alliance out of friendship for dairy farmers or to strengthen any farm organization. Borden's one obsession for forty years had been to buy milk at the lowest possible price and sell to consumers at the highest. They had always opposed and helped defeat dairy farm organizations. After the defeat in 1916 they had threatened to "give farmers enough of organizations." No one knew just what the details of the agreement were, but all careful observers knew that Borden's went into the deal only because they believed they could buy milk cheaper through it than in any other way, and that the arrangement gave them the power to fix the price. Later we shall see that this view was fully justified.

The purpose to break up the unity of the farm organization was deliberate, systematic and ruthless. This purpose was indicated by Borden's cultivation of the League officials from 1916 on. Many farmers in a position to observe it referred to it almost from the beginning and charged that the dealers had acquired influence over the League officials. I had observed it and had cautioned the officials to beware of it. A direct purpose to divide the united farmers was indicated in the first contract made with the League officials in October, 1917, in which farmers of one group were to have a higher price than the others. It was indicated again in the proposal to sell the old Borden plants to the Dairymen's League, Inc. on installment payment and a twenty-five-year option on the milk. It was indicated in practically every important movement affecting milk up to the date of this agreement in the early part of 1922 and since.

Nothing could be more certain at the time of the Borden-League alliance than the purpose to destroy the unity that existed up to that time and to force all producers either into the Borden-League controlled pool, or out of the dairy business. In other words, it was a bold and determined purpose to create a complete monopoly of the dairy business in the New York milk shed, so the first essential was to break up the unity and destroy the resistance of dairy farmers. Borden's did not show their hand openly, but their direction was plainly discernible. Open discussions of the plan of the co-operative, of the "pool" contract and of the procedure were discouraged and largely defeated. No concession whatever was given the producers who felt they wanted to remain in the organization but could not blindly bind themselves in the one-sided "pool" contract. Disputes, ill-will and bitterness were fomented among them. Relatives and friendships of a lifetime were separated and estranged.

Borden's were preparing for the execution of their original threat, which in its purpose and in its execution was sinister and ruthless.

The League officials fell into control of a united organization of the dairy farmers of the whole milk shed. Their management divided it into groups and created disorder, chaos and ruin, but in the face of each new disaster they blamed other producers for their failure, proposed a wonderful new scheme and promised a fabulous dairy future, if the old victims would continue their contributions of milk, funds and credit.

Reunion of Farmers Prevented.

The alliance not only divided the once-united farmers but has persisted successfully for twenty years to keep them apart in spite of many efforts to reunite them and the anxiety of all of them to be reunited.

In 1922 Herbert E. Cook, as president of the New York Dairymen's Association, attempted to reunite them. He told me the League officials defeated his efforts.

In 1924, at a time when the conditions were most ruinous, I gained consent of all five groups existing at the time to come together and work out a plan on which all could agree. We met at Utica and appointed a committee of three from each of the five official groups to develop a plan. The committee held several meetings. A League member was made chairman. The other groups made concessions. It was felt that great progress was being made. Then suddenly, John D. Miller, who had been out of the State, returned and put up an ultimatum which resulted in the withdrawal of the League delegate as he evidently planned or ordered, and, of course, defeated the purpose.

In 1926 a group of farmers in the northern counties renewed the attempt to reunite the dairymen. League officials joined but insisted that no plan should be accepted unless it had unanimous approval of all the five groups, reserving for them the power of veto against all the others. A plan of reunion was solemnly presented by John D. Miller. After more than an hour of reading he was forced to admit that it was practically the Dairymen's League set-up. A plan was worked out but the League delegation voted against it, and the third formal attempt to unite failed.

Then the northern New York producers, of which a major number were League members, said in open meeting that the officials would not agree to a reunion because they knew some of them would lose their positions. So they selected a committee of laymen. Peter G. Ten Eyck was retained as chairman. After working for several months, holding hearings and accumulating information, an agreement was reached. At a large convention in the city of Utica in 1927, an agenda was prepared by a committee headed by John D. Clark, a League director. It directed the officials of all the groups to meet in Albany on a given day and put the agreement into practical operation.

In the meantime, to simplify the problem the non-pool, Eastern States and independent groups requested me to merge all these three groups in one body. I prepared the charter and by-laws of the Unity Dairymen's Co-operative Association, Inc. and filed its charter with the consent and approval of the three groups, so that we had reduced the disordered producers to three incorporated groups and made it easy and practical for all three groups to act as one unit by co-operation among the leaders. All dairymen were eager for it. This last attempt at union was initiated by League producers and developed by lay members of all the groups, no officials taking part in it.

The Sheffield and Unity groups met in Albany on the appointed day and hour. Chairman Ten Eyck was there to call them to order and then leave them to themselves to work out the administration details. While waiting for the League officials to arrive, a message was delivered from them saying that they believed the plan would be illegal and they would not take part in it.

The inevitable conclusion from all this and many other incidents before and since was that the Borden-League combination,

THE BORDEN-LEAGUE ALLIANCE

having broken up the unity of the dairymen as an essential to a complete monopoly for itself, would not consent to a reunion that would defeat their long cherished ambition for a milk monopoly. That combination wanted cleavage not unity. Strife and chaos among producers served its purpose better than peace and order.

Borden's Mark On Everything.

Evidence of Borden's hand runs all through the structure of the Dairymen's League Co-operative Association, Inc. Going back fifty years or more to the first Borden contract with producers we find two pages of legal cap binding the farmer in every particular from the feeding of the cow to the delivery of the milk at the Borden plant. Every act of the producer is regulated and besides there are many acts which he must bind himself not to do. Nothing is neglected. The single mutuality to make the contract legal is in the provision that, if Borden's did not reject the milk but accepted it at the plant and used it, they agreed to pay the price they fixed for themselves.

Compare that contract with the terms Borden's tried to force on the Department of Markets in 1916, but failed to get. Compare it with the contracts Borden's made with the League officials in 1917 and 1918, when the dealers took the volume of milk they wanted and from whom they chose and left the farm-owned plants without a market. Compare it with the "pool" instrument and you find the latter a typical Borden contract, except that it tightens the grip on the farmer's throat with a tenser grasp. The Borden contract did not penalize the farmer for not shipping, or bind him to deliver before he knew what the price would be, or to accept classification, or to pay the cost of distribution, or to make a loan out of the pittance of his return nor to deny him an accounting. The "pool" contract added all these kinks to the original contract but it was nevertheless typical of the Borden instrument. The Co-operative Corporations law, the Milk Control law, the Rogers-Allen law, the Bargaining Agency set-up and the Federal-State Orders might all have been written by the same author. There is no suspicion of mutuality with farmers in any of them. All of them nullify his inherent rights, and fail to give the farmer even a single practical benefit.

Borden's In Court.

We are not without pertinent testimony of Borden's part in the Dairymen's League structure aside from similarity in literary form. In 1924 a law suit was tried at Carmel, Putnam County, New York. One Edward McGuire, a former employee, had sued the Borden Company and its president, Alfred Milbank, for services which he claimed to have performed in negotiating the contract in 1922 between the Dairymen's League Co-operative Association and the Borden Company. His demand was said to be $250,000.

The record of the suit never became available, but some time later I met one of the jurymen who sat on the case. He was a plain, small business man, mild in speech, unemotional yet firm in manner, and rather above the average intelligence. He told me he gave the testimony careful attention because he knew that dairy farmers were "hard up" generally and he thought an organization would be a good thing to help them get more money for their milk. He said Mr. McGuire testified that on a certain day he discussed the matter of a League deal with Mr. Milbank, and that they went to lunch together. At Broadway they were stopped by traffic and while waiting Mr. Milbank agreed to give him $250,000 if he succeeded in making the deal.

In defense, the juror said, Mr. Milbank testified under oath that Mr. McGuire rendered no service in the negotiations in any way. That the League plan was made up by himself and his associates and lawyers in his lawyer's office, and Mr. McGuire had no part in them.

I was not present at that trial and have no corroboration of the information. I do not know whether Judge Seeger of Newburgh, Orange County, who tried the case, dismissed the case on a point of law or that a compromise was made, but the case did not go to the jury. No appeal was taken. McGuire promptly disappeared; it is said that he went West.

There is significance in a statement contributed by Mr. R. D.

Cooper to the Utica Press published April 29, 1924. In the press release among various things Mr. Cooper said:

Supposed to be a Secret

"I understand that it is a matter of court record that McGuire, formerly on the payroll of the Borden companies, who brought suit first against the Borden companies and later against Mr. Milbank for $250,000 for services rendered in negotiations with the Dairymen's League, testified that in December, 1920, previous to the election of the League officers, he told Mr. Milbank that John D. Miller would be elected president, that Miller would resign and Slocum would be elected president." (This prediction is just what did happen.)

"The complaint in the case recites conferences with League officials and representatives and communications of which the Executive Committee had no knowledge nor ever any report." (Mr. Cooper was chairman of the Executive Committee.)

"It was supposed to be a secret but the Borden Company was given rebates on milk shipped from League plants of as much as 20 cents per hundred, which amounted to hundreds of thousands of dollars."

(Signed) R. D. COOPER

There have been direct reports in the past of rebate checks up to $50,000 at a time from the League to Borden's. More recently it has been reported that Borden's benefits in the League preference to them amounted to fully $200,000 a month or $2,400,000 a year.

During State Control it was brought out by Henry Manley, attorney for the State, in Court in the City of New York, that the League bought milk in the State of Pennsylvania at eighty-six cents a cwt. below the New York price and sold it to a Borden subsidiary at a cut price in competition with New York milk.

The Federal Trade Commission reported that correspondence between Borden's and the League showed that the League consented to Borden's demand for a reduction on milk which the League shipped from Pennsylvania and delivered to Borden's. The audit of Dairymen's League books by Ernst & Ernst on behalf of the Commissioner of Agriculture and Markets revealed an outlet of $6,055,167.43 for the fiscal year 1937 which the League management was requested to explain. The explanation was that this was a "price adjustment which the association was compelled to make to its customers." The records did not show how much of this $6,055,167.43 was refunded to Borden's. No allocation of it to "customers" was available. But Borden's take about one half of the milk handled by Dairymen's League, so it is safe to conclude that as one of the customers, Borden's got at least their share of the rebates.

CHAPTER XXIV

SHEFFIELD FARMS STOOD ALOOF

Sheffield Resisted.

When the split-up took place in 1922, Loton Horton was president of Sheffield Farms. He had no illusions about Borden's position or purpose. He knew right well that Borden's would be the master of the Borden-League alliance. He knew that to join it would be to subordinate his company to Borden's domination. He refused to go along. He, however, did not join the non-poolers but organized a producers' association to supply Sheffield Farms exclusively. It took in many of the outstanding dairymen of the State. They included Fred Boshart, of Lowville; Hugh Adair, of Delhi; Homer Jones, of Homer; J. Leslie Craig, of Canastota; Clark Halliday, of Chatham, N. Y.; Fred E. Mather, of Ulster, Pa. and many others of their type of producers. The State could boast of no better citizens.

These men were not "kidding" themselves. They knew they were not organizing a real co-operative association. They rebelled against the high-handed way their interest in the old League had been appropriated and squandered. They abhorred the "pooling" contract, and the "pool" set-up. They realized from the start that such a combination would "have dairy farmers by the throat" and said so in those exact words. They did not object to the pooling of milk, no group of dairymen did, but they balked at the contract, at the arbitrary and autocratic official control of the association, and at Borden's power to fix milk prices for themselves and for others.

Sheffield Farms continued to announce a flat price in advance and up to 1933 paid regularly month after month from thirty cents to one dollar per cwt. more than the Borden-League combination paid. While the Sheffield Farms' producers formed a separate unit, they co-operated in every attempt of the other independent groups to reunite all producers. They knew that with Borden's getting their full supply at a price fixed by themselves, such a volume would fix the price for the metropolitan market and basically for the whole state, and all any other dealer could do was to economize on distribution and share the savings with producers.

Everybody in the trade believed that Borden's would get a price concession, but no one could prove it because the League producers were barred by the contract from demanding an accounting. It soon became evident, however, that it was costing millions to support the complicated and extravagant League organization and its internal politics. The cost of the Sheffield producers' organization was negligible so that Sheffield Farms could pay a good bonus over the League and yet stand pat with Borden's on city prices to consumers.

Borden's and the League realized the strength of Sheffield Farms position and soon made a desperate attempt to get Sheffield Farms into the combination. Their leaders spent days and nights in the homes of the Sheffield Producers' leaders in an attempt to persuade the producers to line up with the Borden-League alliance. They failed. One inducement after another was offered Loton Horton. He sternly said, "No." Shortly after this experience Sheffield Farms Company became a subsidiary of the National Dairy Products Corporation.

The League's Provisional Contract.

Still persistent in its ambition for a monopoly in New York, the Borden-League alliance developed a scheme in 1932 to win enough Sheffield and independent producers into their clutches to break down the resistance of Sheffield Farms and the independent producers. What was alleged to be a spontaneous demand of dairymen for one organization turned out to be largely a selected group who voted unanimously that the one organization must be the Dairymen's League. A provisional contract was devised. The dairymen were required to sign the "pooling" contract with the provision that it would not be binding unless 70 per cent of Sheffield producers and 75 per cent of independent producers of other groups signed before September, 1932.

The campaign for signers to this "provisional contract" was carried on for five months. Meetings were held all over the State. The issue was discussed pro and con. Educational influences at Cornell and St. Lawrence Universities and elsewhere were utilized in favor of the scheme. When the contracts were counted they fell far short of the quota required. The drive failed. The cost of it to League producers has never been made known.

The League official announcement following the failure of its transparent scheme was typical. Farmers who refused to step into their trap were told that the League had adopted a new policy of restricted membership. In other words, they could not now get in. The League had become a "closed shop."

I suppose that Sheffield Farms acted in all this for their own interests. Their producers were also working properly for their own interests, but I knew them intimately and I know they realized as I did that no group of producers could benefit permanently in a plan that did not comprehend the welfare of all producers.

If in those critical days Sheffield Farms Company had adopted the "pool" contract and required its producers to sign the contract, a large number of our best dairymen would have been forced to sell their cows. They just would not sign the "pool" contract. Many, of course, would yield rather than lose their homes and farms that were equipped for dairying. The dairy business would be a closed monopoly and without the intervention of a miracle our dairy farmers could not escape a state of serfdom. So no matter what has happened recently or may happen again, and without regard to the incentive, Sheffield Farms did New York dairymen a good turn when help was needed.

Farmers Not Members.

The publication in *The Rural New-Yorker* in 1921 of the fact that the membership of the Dairymen's League Co-operative Association, Inc. was limited to twenty-four members, who were the twenty-four directors, and that producers had no part in it stirred farmers to a spontaneous protest. One of the directors said, "Dillon raised hell with us for a time." They, however, put their entire force in the field to tell dairymen that Dillon was "an old grouch who never agreed with anyone unless he had his way about everything. He was sore because they would not let him be president of the League. He was opposed to co-operation and an enemy of the farmer."

Of course, if all that and more that they said were true, it would be no answer to their deceit and breach of trust. For the time being the trick satisfied some. It satisfied others for a while, many it never satisfied for a moment. But they were in the scheme and had nowhere to turn if they bolted. The officers attempted to brazen it out and did so for a while. But George Holmes and Merle Holmes of Westmoreland, Oneida County, N. Y., accepted the "pooling" contract in 1920 and continued shipments to Clover Farms as before up to May 1, 1921. From that date Clover Farms paid the co-operative for Holmes' milk and it paid Holmes Brothers. On April 1, 1922, Holmes Brothers discovered that Clover Farms had paid the new League during the 11 months $4,207.77 of which only $3,015.11 had been paid over to the producers, leaving deductions of $1,192.66, or more than one-fourth of the product to the League. Holmes Bros. stopped deliveries. The League officials held up their last month's milk bill and brought a suit in the Supreme Court for "liquidated damages" as fixed in the contract. The trial judge held that all dairymen who signed the "pooling" contract were members and that the liquidating damages were legal. The jury was divided but on these instructions from the Court, the jury found for the plaintiff.

Holmes Bros. appealed to the Appellate Division. The case was argued in September, 1923. The decision of the trial court was reversed and a judgment rendered in favor of Holmes Bros. for the amounts due with interest and costs. The Appellate Division held

1. that Holmes Bros. were not members of the Association, and that the plaintiff did not contend in court that they were; and could not do so in view of the evidence; 2. that the deductions for capital loans, etc., were clearly unlawful; 3. that the deduction

for a subscription to the *Dairymen's League News* was without authority of law; 4. that "liquidated" damages must be reasonable in amount and fairly relative to damages ordinarily suffered in like circumstances.

In other words the provision for "liquidated" damages fixed in the "pool" contract may be unlawful. Damages must be proved. The only effect of the clause in the contract is to cause farmers to fear heavy penalties for discontinuing deliveries of milk, and furnishing an excuse to the League for not settling with farmers for milk bills due them.

The League officials carried the case to the Court of Appeals, the last resort in the State. The judgment of the Appellate Division was affirmed.

There was at the time a large number of similar claims by farmers who had, like Holmes Bros., discontinued deliveries and had not been paid for milk delivered before withdrawal. These claims were then presented and paid.

The opinion is held by many laymen and lawyers that dairy farmers are complying with many provisions of the "pooling" contract and dairy statutes which would be held illegal if tested in the courts. The co-operative terminates its contract on April 1st if the withdrawal notice is filed. This is the hardest time in the year to find a buyer. Many patrons have stopped deliveries at other dates and the amounts due them at the time are not paid, though the farmer be willing to allow the damages permitted by law. Consequently, the co-operative has accumulated a large sum of money that rightly belongs to producers.

CHAPTER XXV

LAYING THE BASIS OF MONOPOLY

League Plan Failed Again.

The new League first launched out on a plan to manufacture by-products of milk. It bought and equipped a number of expensive manufacturing plants. President Slocum went to California and hired a salesman on a five-year contract at $25,000 a year. An advertising program was put on at a cost estimated at $5,000,-000. We read propaganda about the wonderful success in an ice cream and condensed milk venture, not only in New York but around the world.

Later there came whispers of Borden's objection to the distribution of ice cream in their territory. After about four years they admitted that their plans were a great success but some "slight changes" were necessary. *Hoard's Dairyman* pricked the bubble of pretense by saying that if it were a "great success" the plan would not be changed. The "slight changes" consisted in selling the manufacturing plants that Borden's wanted to that concern and turning the manufacturing business over to the Borden Company.

At best the scheme was contrary to the best interests of dairymen. Farmers received their highest prices for fluid milk. They received a much lower price for milk to make condensed milk. The League had the agency from farmers to sell their milk at the best price obtainable and here that agency was using the farmers' money to induce consumers to stop buying fluid milk or to curtail their purchase of it, and substitute condensed milk, which was produced from milk that farmers were forced to sell at a surplus price. Farmers' money paid the advertising bills to defeat their own best interests.

League Becomes a Holding Company.

At the time the "pool" was organized farmers owned a large number of country plants. They had come to realize that ownership of the plants gave them power in the negotiation of prices. The Borden-League combination determined to control producers by gaining control of farm-owned plants. It used capital money deducted from all its producers to buy up the farm-owned plants. Many of the farm groups sold or rented their plants because of their faith in the organization. Most of them have regretted the sale since. In the meantime the alliance began to buy out the large independent dealers who had plants in the country and competed with the Borden Company in the city.

The announcement was made that the League had bought out the Empire Dairy Company. After some time another announcement was made that the League had sold the city end of the Empire Company business to the Borden Company. This was repeated in several purchases until the Borden competitors had practically disappeared. To ward off criticism, the League management announced that its permanent policy was to favor dealers buying League milk. Then the League began to buy out dealers supplying can milk to stores, hotels and other milk buyers of that class. Several of these had country plants. Some of them were continued by the League in their own names as subsidiaries. In this way the organization became a holding company and a milk dealer itself, as well as an agent for the sale of farmers' milk.

In some centers there were two and sometimes three plants owned by different dealers. As the League gained control of them it closed and usually dismantled or demolished the entire buildings. It would sell the plants for other purposes but would not allow them to be used again as milk plants either by the local farmers or by other dealers. In territory so controlled the individual farmer has no independent choice. He must turn his milk over to the Borden-League alliance or sell his cows and go out of the dairy business. The combined leaders boasted of farm loyalty for political effect but they did not trust to it. They took no chances on sentiment or friendship. They made sure of economic power over individual farmers. They won patronage over the helpless but they have lost what is held more precious by most men—confidence.

The League practice was to deny all rumors of a purchase and sale of an independent plant until a day or two before the transfer. Then they notified farmers that they must sign the "pool" contract before the milk could be delivered at the plant. Canvassers were sent to the area to induce patrons to sign the contract. The procedure provoked the resentment of dairymen. At first it was easy for farmers to divert the milk to other nearby plants. But when such plants had been bought up and closed, farmers were obliged to hire a truck and cart the milk ten or fifteen miles to distant plants. In other places the farmers gained possession of old buildings and hastily equipped them to handle their milk.

The Clover Farms plant at Homer, N. Y. was receiving about 650 cans of milk daily. The League bought it and gave two days' notice. The farmers acted promptly. They gained possession of an idle building, interested Sheffield Farms in it, and shifted their milk to it under temporary equipment, which was later developed into a high class permanent plant. The strenuous canvass of producers succeeded in saving only a few local patrons. It was necessary to ship milk in from distant fields to keep the Borden-League plant open. But in most cases the buying up and closing of country plants left many farmers no choice but to submit to the monopoly and turn their milk over to the Dairymen's League, the only plant within their reach.

Borden's Makes Sure of Surplus.

The policy of the League to gain possession of plants and milk carried them into sections of New York, Pennsylvania and other States which had not shipped fluid milk. It converted sections that previously made butter and cheese into shipping territory. In the Champlain section in the north-eastern part of the State it carried on a campaign for members and built plants at a cost to farmers of about $1,000,000. This brought trainloads to the city.

At this time they worked up a scare of the "western menace," as they called it, insisting that we must have more milk to keep western milk out to save our markets. It was not long after that they complained that the surplus of milk caused low prices. Surplus was always an asset to milk dealers. The profits in its manufacture up to 108 per cent will be shown later.

The building of our system of hard surface roads and the perfection of the auto-trucks have facilitated the assembling of milk in central country plants. This has reduced the cost of handling milk in the plants and the tank truck cartage has reduced the freight rate more than one-half. The buying out of competitive plants and dealers has reduced competition of the dealers in both the country and the city, but the farmer has not profited from any of these changes. As a whole he has suffered from them.

The reduction of freight charges of more than one-half went wholly to the dealer. Some farmers may get a little of the savings of the centralized plants, but they are obliged to pay it back for trucking long distances to the plants. The closing of their local plants deprives them of competitive buyers and heaps the trucking charges on their backs. When patrons of selected plants get a bonus, patrons of another plant get that much less.

Plight of Dairy Farmers.

The plight of the dairy farmer had become so serious by the winter of 1932 that a legislative commission was appointed to inquire into the cause of the spread between producers and consumers. Senator Perley Pitcher of Watertown was chairman of the commission, James Cross was selected as counsel and Dr. Leland Spencer of Cornell was engaged as investigator. Dairymen's League officials were not keen about the investigation but they had a safe majority of the members of the commission. Its counsel was decidedly pro-League. It held hearings in many parts of the State. League attorneys, officers and representatives were generally present. Farmers attending the meetings reported that any testimony pointing to the responsibility of the League for low prices or other troubles was challenged or minimized.

During the 1933 session a bill was introduced to create a Milk

Control Board with authority to investigate the production and distribution of milk in all its phases, to license milk dealers, and, as an emergency measure, to fix the price to be paid farmers for milk for one year. The League management opposed this bill openly and strenuously. Sheffield Farms Company and other milk dealers also opposed it.

Holding that it was the inherent right of dairy farmers to negotiate for themselves the price of the milk they produced, that they had been deprived of that right by a combination of co-operative leaders and milk dealers, and that it was the duty of the State to restore these rights to milk producers, I had favored an investigation in the hope that it would reveal not only the plight of dairy farmers but also the real cause of it, and a remedy for it, but I was opposed to the usurpation of farmers' rights by any State bureau whatever. I knew, however, that farmers were hard pressed financially and when a hearing was promised on the bill I wrote a large number of milk producers in some twenty-odd counties of the State simply saying I would attend the hearing and asking them, after all the discussion, what they wanted me to do or say on their behalf.

The replies were practically all in the same form and tone. The dairymen wrote that they knew what I had said was true; they were familiar with my position and principles and approved them and they would support them always. But they wrote, "We are just now in desperate straits. Our taxes and interest on the mortgages are past due. Our feed bills are unpaid. Our credit is exhausted. We have no money to buy seed or other essential supplies and the spring season is near. Every can of milk we sell leaves us further in debt than we were before we produced it. This robbery must stop soon or reform will be too late to help us. Get us out of this hole now and we will help you fight for sound principles and our rights as soon as this crisis is over."

There were replies from some farmers who had reserves to cover their current milk losses, but their reports of conditions in their neighborhoods were the same as from the distressed farmers themselves. These replies come from all groups in the State, including producers for the League and Sheffield Farms.

At the hearing a humorous incident was noted. The League officials usually have a delegation massed and badged to applaud or decry a policy as discussed. When I rose to speak these League delegates were generous in applause, expecting opposition to the bill. I briefly restated my well-known principles and related the result of my correspondence with dairymen. Under such conditions I said I would approve the emergency bill provided its proponents would help during the emergency period to revise and perfect our dairy co-operative system under which all milk producers could unite, so that when the emergency was over we would have a co-operative system ready to carry on without jolt or confusion. The League delegation did not join the proponents of the emergency bill in approval of my suggestion. But the approval of the bill was so overwhelming it was felt that all opposition had been overcome because of the need of prompt action.

Dealers Write New Bill.

The next day, however, the opposing leaders and dealers assembled and worked out a substitute bill of their own. Instead of the minimum price to be paid farmers, the new bill provided for a maximum price for dealers to charge one another, stores and consumers. It also provided that "unadvertised brands," meaning practically all except Borden's, the League's and Sheffield's, might be sold at one cent less than the price fixed for "advertised brands." This was first suggested by independent dealers. The big dealers promptly approved because of the extra cent they might charge on their family routes. Later Borden's contested this provision in the courts, which sustained the law, but its purpose was defeated by subsidiaries of the big dealers and finally by a general disregard of the Control law.

The ban on loose milk, which went into effect July 1, 1933, was the real reason for the higher price to consumers on so-called "advertised brands." The price to the farmer was the same and the quality was the same. But now all milk for families was to be bottled and it was feared that the full extra cost of bottled milk would reduce consumption among consumers who previously paid less for loose milk at stores.

The provision in the original bill to fix minimum prices to be paid farmers was omitted as planned. The theory was advanced that fixing the price to consumers would stabilize the markets and then the dealers would generously pay farmers a fair price. This abandonment of the farmer, whose plight started the whole show, created opposition. Senator Byrne, who was at the time Chairman of the Agriculture Committee of the Senate, insisted on some direct protection for the farmer. The dealers' bill was then amended to provide that, in the event that the dealers did not make fair returns to producers, the Control Board might hold hearings and fix a price for producers.

The bill as passed contained a further provision which exempted co-operative associations of the Dairymen's League type from the provision fixing the price for farmers. While it did not mention the Dairymen's League by name, the intention was clear that all other dealers and co-operatives were required to pay farmers a definite price, if and when fixed by the State while the League might continue to pay its producers what it and Borden's pleased. "Classification" and "blended prices," the twin enemies of producers, were also for the first time legally sanctioned by law. The co-operative exemption clause had not been in the original bill. I have never been able to find anyone, legislator or layman, willing to admit authorship of it, or who admitted that he knew when the "exemption" clause was inserted.

Governor Lehman had let it be known that he was in doubt about approving the bill if passed. But in the meantime farmers all over the State protested the change in the original bill and the delay in any relief from a maximum of two cents and as low as one and one-quarter cents a quart for milk. A strike developed in western New York. The following dispatch from Rochester indicated the situation:

"But during the week thousands of gallons of milk were dumped into the streams and gutters. Dairymen's League trucks persisted in bringing milk into the city. There were conflicts between the striking farmers and the troopers who were armed with helmets, tear gas, masks, clubs and some with guns. There were bruises and sore heads on both sides. At the Monroe-Wayne County line 400 dairymen obstructed a train of

LAYING THE BASIS OF MONOPOLY 207

10 trucks from other counties and had a conflict with the troopers. There were many sore heads, but every quart of milk was dumped.

"President Woodhead of the Western New York Producers' Association predicted that the Pitcher bill would become a law and with a price fixed at 3½ cents a quart the strike would end in success. Producers who backed the League's strike of 1916 and also of 1919, when prices were much higher than now, are chagrined that the management of the organization is now trying to defeat them in a fight to gain a smaller price in their own local markets."

With this threat of a disorder all over the State, Governor Lehman sent an emergency message to the Legislature and the bill passed promptly. The law created a milk control board of three members. The governor appointed Charles H. Baldwin, Commissioner of Agriculture, Dr. Thomas Parran, Jr., Commissioner of Health, and Kenneth Fee.

It was not until the Pitcher bill became a law that farmers realized it had been changed so that the Control Board was mandated to fix prices to consumers for the comfort of dealers but not required to fix a price to producers; on the theory that if a satisfactory price was stabilized in the city the dealers would then of their own volition pay producers a reasonable price. Farmers said openly that they were tricked. Their anger rose by the minute. They demanded a hearing before the Control Board. It was the most stormy farm meeting on record at Albany. The Board had increased the dealers' price to consumers one cent a quart and left the price to producers to the tender mercy of the dealers. After a hearing, the board fixed $1.88 per hundred pounds to farmers for fluid milk. This meant about a half cent a quart increase to producers in the blended price.

Disillusioned and desperate, dairy farmers rose in a wide area in protest and rebellion. A strike was called, milk plants were closed and shipments were suspended in many sections. Disorder became rife and the Governor sent armed troopers to the disturbed areas. Clashes occurred, guns were used by the troopers and for a time disturbances increased. Finally Governor Lehman directed the Control Board to fix a minimum price to be paid farmers and promised another investigation of the dealers' spread, and with these concessions induced the striking farmers to suspend the warfare pending investigation and correction of abuses.

Dr. Spencer spent $50,000 in his investigation for the Pitcher Committee. He submitted figures only of groups and reported that many dealers were operating at a loss.

Governor Lehman sent Dr. Spencer to New York a second time to investigate the spread. This time the doctor had $50,000 more, which he reported altogether insufficient, but he was able to report that dealers had to distribute 143 quarts of milk to make one cent profit. The Federal Trade Commission reported that for that year Borden's profits were $7,051,872, after paying all salaries, taxes and expenses.

CHAPTER XXVI

STATE MILK CONTROL

The League Officials Cashed In.

For the first three or four months the dealers quite generally seemed to comply with the Control law. Then they woke up to the significance of the Dairymen's League exemption clause in the law. They were paying producers up to a dollar a 100 pounds of milk more than the League was paying its producers in the same territory. The League was the third biggest milk distributor in the State. With this saving in cost, the League management cut prices in the city below the State prices. In at least one instance it made a loan to a chain of restaurants in an amount said to be $75,000 to win the trade away from a dealer paying the full State price to producers.

Muller Dairies, a National Dairy subsidiary, had four plants, one in Ulster County and three in Madison County. During one month of 1934 this dealer reported that it paid its producers $1.03 per 100 pounds more than the League paid its producers. In the city it was in competition with the League as dealer. It filed more than 200 complaints in violation of the Control law against the League in the State's New York City office. Burt Miller, a former employee of the League, was the head of that office. Mr. Miller prosecuted none of them, giving as his excuse that convictions could not be secured in any event because the storekeepers would deny that they paid less than the Control price. In other words he contended that offenders could not be convicted unless they confessed guilt.

Muller Dairies expressed its willingness to continue the full Control price to producers provided the Control Board would enforce its retail prices in New York. This the Board would not guarantee to do. But Commissioner Charles H. Baldwin did breach the law to allow Muller Dairies to pay its producers less than the Board price to producers. This arrangement continued until Commissioner Baldwin was succeeded by Commissioner Peter G. Ten Eyck, who refused to continue the concession. Muller Dairies then turned its Madison County plants over to the Keystone Dairy Company, another National Dairy Products Corporation subsidiary, which reported uses of the milk largely in lower price classes. The Ulster County plant was bought by the Dairymen's League but the patrons dropped out and the plant was closed. When Muller Dairies gave up its plants, it bought its supply from the Dairymen's League, indicating that it could buy fluid milk cheaper from the League than from farmers direct.

Later, sleuths of the racketeers allured the producers of the three Madison County Muller Dairies plants into a dealer-controlled co-operative sham, and the milk of the three plants fell into the control of the Borden-League combine at prices to producers far below the State Control prices. The State law exempted the racketeers and legalized the racket. The State was a party to the deception and primarily responsible for the farmers' losses.

How Classification Works.

After complying with the Control law for some months Sheffield Producers' Co-operative Association made contracts with its individual members and changed its legal framework otherwise to qualify for the co-operative exemption in the Control law. This had the effect of reducing the returns to Sheffield producers. Sheffield Farms Company continued to pay its producers more than the League management paid its producers, but not as much more as previously. The average margin was reduced still further in 1937 when it joined the monopoly.

There followed the most hectic conditions that I have known in the milk business. The Control authorities fixed a high price for Class 1 or fluid milk to city consumers and a comparatively low price for manufacturing purposes. The Class 1 price varied from $2.305 in January, 1934 to $2.175 in June, 1934. It was raised to $2.45 on June 11, and from July, 1934 to August 15,

STATE MILK CONTROL 211

1936 the price continued at $2.45. From August 16 to September 25, 1936 it was $2.70 and from the latter date to March 31, 1937 the Class 1 price was $2.90. The prices fixed for consumers for these periods were:

Year	Average price per quart
1934	12.3 cents = 5.781 per cwt.
1935	12.5 cents = 5.875 per cwt.
1936	13.1 cents = 6.157 per cwt.
1937	12.6 cents = 5.922 per cwt.

This suited Borden's and Sheffields, both of whom had a large fluid family trade, and the margin of profits was high. When a family reduced its consumption of fluid milk it would use condensed or evaporated or powdered milk instead. The price of the milk used to make these products was low. It was called "surplus." The price of these products was and is based on the prices fixed for cheese at the Plymouth Exchange in Wisconsin and on the price of butter and condensed milk fixed in Chicago. The Federal Trade Commission reports tell us that the price of these commodities is fixed in sham sales and formulae by a few big dealers, headed by Borden's and the National Dairy Corporation. In consequence, while the dairy farmer was led to believe that State control and "classification" had increased his fluid milk price, the system reduced the volume of his fluid trade, and increased the volume of surplus, which the dealers always allege is the cause of low prices for all milk. So the farmer got less and as the records show, the dealer made bigger profits on the so-called "surplus" than on the fluid milk.

This is not all. According to the Federal Trade Commission, some of the same dealers operating in Philadelphia and Connecticut sold fluid milk at Class 1 prices and paid farmers for it at the lowest "surplus" classifications. The Commission report calls it "cheating farmers out of large sums of money."

New York farmers were told that audits were made to see that they got the Class 1 price for all milk sold for fluid use. It is admitted now that no such audit was ever made. This is discussed later in more detail in connection wtih Attorney General Bennett's Report. In other words, dealers were free to make the blended price to suit themselves.

The Control Board also fixed the price for the city stores as well as for hotels, restaurants and other like distributors. While Milk Control was promised to reduce the dealers' spread, the Attorney General's report shows that the spread increased nearly a cent (nine-tenths of one cent) during the period of State Control.

Effect of the League's Exemption.

The Dairymen's League with its exemption privileges had at the beginning a big margin in price below the dealers who attempted to comply with the Control price to farmers. It and its subsidiaries cut the price to stores to win trade from other dealers. They did not wait long for the reaction. In the scramble that followed, stores were buying milk for from five cents to six cents a quart and selling at ten cents to eleven cents. Their normal profit is one cent a quart. Sometimes they were arrested and fined. They paid the penalty and went right on. They were making big profits but they did not like it. They frankly told Senator Rogers' 1937 investigating committee just what they did and how they violated the law. They had no choice except to go out of business. They told Mr. Rogers they wanted him to stop the biggest racket they had ever known in the milk business. Many dealers who had plants in the country saw that they could not pay the Control price and compete with Borden's, the League and Sheffield's, who were authorized by the Control law to pay less. The Big-3 would soon have all the trade and their business. The small dealers had to have a cheaper price or close the country plants.

The producers saw the situation. For them it meant helping their buyers to keep on or being forced into the big co-operatives or going out of the milk business. They preferred to save their local markets. The dealers organized co-operatives so as to get the exemption. Co-operation, which was originally devised to enable dairymen to negotiate the sale of their own milk and to increase their price, had been so bedeviled by chicanery and discriminating law that as a matter of actual practice it had forced farmers to reduce the price to themselves. They fully realized that to surrender to the racketeers of the monopoly would mean not only a permanently lower price for the future but also the total loss of their economic freedom. They also began to realize that without economic freedom, their civil freedom could not long survive.

State Encouraged Perjury.

A few dealers boldly announced that they could not get licenses without perjuring themselves, which they refused to do, and they could not pay the Control price under conditions existing. Hence they proposed to handle milk as "bootleggers" and would pay their producers as much as other dealers paid, provided the producers continued to deliver milk on such terms. Some did. Other dealers made a contract to sell their plants to producers at fabulously fictitious prices and the producers agreed to make monthly payments on the contracts. The payments were deducted from the milk returns figured at the contract prices and so reported to the Control Board by producers and dealers.

There were a few genuine co-operatives which sold direct to city distributors. These sales were made largely through milk agents. In all these cases the dealers and the co-operatives were required to make sworn statements to the Control Board monthly, and they were all obliged to pay and charge Control prices or be prosecuted for violation of the law and subject to fine and loss of license. In the city the dealer sold to the store at the Control price, collected his bill and paid back a rebate. When a country plant or a co-operative sold to a city distributor, the contract called for the Control price. This was paid to the agent and a rebate was returned to the buyer.

This was the system that the Emergency Milk Control law forced on the dairy industry of New York State. There were honest dealers who revolted against it. There were honest farmers who were distressed and bewildered with it all. Under ordinary conditions the sworn reports in essence would be rank perjury, but under the circumstances they were formalities complied with under duress. People must live. They are under obligations to support their wives and children. No individual or small group can resist the power of a whole State. The moral turpitude was not in the minds of the victims. It was in the acts of the responsible rulers of the State.

State Authorities Fail.

The most regrettable thing about it all was the bad example for the young men and women of country and city; especially for those who were just beginning to assume the responsibility of self-support. The State created and renewed a system for four years that encouraged trickery, deceit and crime. It was a reproof to the authorities and a scandal to the State.

This condition of affairs was known to the Control authorities, to the Legislature and to the Governor. Yet it was renewed three times with no attempt to correct its faults. On the contrary, every change made imposed a new restriction on the inherent rights of dairy farmers, and a new favor to the milk monopoly.

During the first two years of his administration, Governor Lehman complained that he had no authority over the Agricultural Department because it was controlled by the Agricultural Council, nine of the ten members being appointed by the Legislature, who in turn appointed the Commissioner of Agriculture. The Governor promised that if the law was changed to return the Department to the control of the Governor he would correct the abuses in the dairy industry. The law was so changed.

He appointed Peter G. Ten Eyck to replace Commissioner Charles H. Baldwin, and again promised to solve the dairy problem. He renewed the Control law just as it was and added three or four more laws, each one of which cramped the milk producer a little more and added a new advantage to distributing dealers. In addition he first encouraged the scheme proposed by selfish dealers and leaders to create a Federal bureau with power to control the flow of milk to New York markets and to regulate prices. The spontaneous and emphatic protest of New York dairymen defeated that scheme.

Next, the Governor allowed himself to be led into a similar scheme known as the Seven-States pact. By this time dairy farmers had come to understand the purpose and effects of the plan even better than before and it was defeated by their unqualified disapproval. To allure additional floods of perishable goods to a market already over-supplied serves the dealer-distributor because it creates a surplus market, but it is cruel and outrageous to both the new and the old victim.

Attorney Protests the System.

On August 8, 1934, Henry S. Manley, counsel to the Milk Control Board, spoke at the Caledonia Fair in Western New York. His speech was epitomized at the time in the following brief paragraph:

"(1) League officials opposed and defeated the payment of producers for their milk twice a month; (2) The real reason for this opposition is that it gives the League $5,000,000 of its producers' money to use as extra capital; (3) The League uses the money to operate its subsidiary companies at a loss and to finance chains of restaurants; (4) The League officials helped other dealers to oppose the enactment of the milk control law; (5) That after their opposition failed, they succeeded in getting an 'exemption' in the law which 'terribly weakened' it; and (6) Ever since, League officers have tried to undermine the law and those who are enforcing it."

At that time Mr. Manley was quite hopeful of the success of Milk Control and the future of the milk industry in New York State. Two years later, in September, he resigned his position to take effect on September 15, 1936. The following excerpts from his radio address explaining his reasons for resigning his position reveal some conditions under Milk Control:

"The reason for my revulsion against milk price control is this: For the past two years it has been increasingly my business to prosecute suits against small milk dealers and against small groups of producers for price-cutting, when I know in my heart that conditions have become so bad that they are helpless to do otherwise. As the attorney for milk control I have not chosen the people I have sued; it has been my duty to defend all suits that were brought against milk control and to sue whatever dealers or producer groups I have been directed to sue. I cannot feel satisfied conducting a lawsuit against some people for doing what I know nearly all others are doing, unless I can see that the situation is getting better. Instead it has constantly been getting worse.

"Anyone with a stub of lead pencil and the figures which are published monthly in cooperative papers or posted at the milk plants can demonstrate that the very largest groups of producers, handling two-thirds of the milk in New York State, are not receiving within 20 cents per hundredweight of what they should receive, according to what the dealers themselves say about the classification of their milk. This fact has not been a secret.

"John Dillon has long been attacking the situation in the columns of *The Rural New-Yorker*. Dairy statistics sent out from Cornell University have several times called attention, rather timidly, to the fact that prices actually received by farmers are less than the prices established by the milk control orders.

"There are important groups of New York State dairy farmers receiving little more than the Mid-West condensary price for milk, and I have shown where some of the shortage goes.

"Short payments to farmers are not all retained by dealers as added profit. The reduced price to the farmer is only half of a vicious circle, the other half is cut-prices to storekeepers, discounts to favored customers, commissions to milk brokers and milk-wagon drivers and solicitors, and other competitive expenses.

"I have told you that my work in recent months has been constantly to prosecute helpless small dealers and producer groups for doing just what everybody around them is doing. It is impossible for me to bring any moral fervor, or even a clear conscience, to that work, and so I am quitting it."

CHAPTER XXVII

GOVERNOR VETOES FARM BILL

Cost of Production Proposed.

At frequent times during a period of fifteen years dairy farmers individually and in groups appealed to me to call a strike again to fix prices for milk. I tried to explain to them that under existing conditions a strike could not win their objective. The reasons are these: The League producers were under contract. Some of them would strike. A major number wanted cost of production and a reasonable profit, but a strike would give them no such assurance. Sheffield producers had a better price and market than the League producers and would be little inclined to join the independents and assume the losses of a strike.

This would mean that Borden's and Sheffield, and the League, as a dealer, would have a large supply from the New York milk shed. All of this milk could be sold as fluid milk. The Big-3 could and probably would increase the price to producers and consumers temporarily to defeat the strike. Each of them had a chain of subsidiaries and affiliates that they could call on for extra supplies from other States. The producers who held their milk out of a market would be hurting their former distributors or driving these men out of business entirely, and destroying their own hope of a future independent market.

Besides all this, I reminded them that the money lost in a strike, if prudently used in acquiring plants and distribution equipment, would put them in a position to command the markets for themselves through a real co-operation. This would again unite all dairymen. Most producers saw the merit of this simple business proposition.

It is true that during the past fifteen years several strikes have served to focus attention on dairy farmers' protests against abuses. This served a good purpose but it is not certain that the producers who held their milk and took the brunt of the fight ever received benefits enough to cover losses, except in the fight of 1916 when farmers were united. At that time practically all farmers were free of contractual obligations. The milk was safely sold to dealers who delivered it themselves to consumers, and a powerful arm of the State (the Department of Markets) directed the sale with the sympathetic support of the Attorney General's office of the State. The Dairy Farmers' Union strike of 1937, discussed later, may be another exception.

The provocation for this was presented by the Sheffield Farms Company after a long period of mutually favorable farm sentiment. The Union was successful in diverting the milk to other plants for shipping or manufacturing and in that way minimizing the loss. The milk, however, was largely used for making cheese. The Kraft-Phenix Cheese Corporation, a subsidiary of National Dairy, was the chief buyer of the milk. Hence the profits went to National Dairy just as they would if the milk had gone into Sheffield's plants. But the strike clearly had a favorable effect on prices. It is too early yet to judge of its permanent results, but it was a timely asylum for northern New York farmers. As such in my judgment it merited support, and it is hoped that the Union will dovetail with other farm movements and be a factor in the final solution of the milk problem.

Another Move for Unity.

During the Fall campaign in 1934, when the Republican Committee was putting up a fight to regain control of the Assembly, a high leader in the committee told me that if a representative committee of New York dairymen would formulate a definite policy for the distribution of milk, the committee would adopt it as a party program. At my suggestion a group of practical dairy farmers, fairly representative of the industry and in position to know the sentiments of farmers generally, met in the late summer of 1935 in Albany and prepared the following statement:

Principles Outlined

"We favor the establishment of an organization for the sale and distribution of milk and its by-products under a State cooperative law which will fairly safeguard the welfare of the whole industry. To this end, the system must be comprehensive enough:

1. To serve equally the interests of every group and every individual dairyman in the State:
2. To provide conveniences and regulations by which every member will have a voice in fixing the general policies of the organization:
3. To provide full and complete accounting and information so that the membership control may be exercised intelligently and efficiently:
4. To authorize and empower its duly elected representatives to determine without discrimination or favor, a fair basic price to be paid producers for milk:
5. To set up an organization, on an equitable basis, of dairy farmers at each shipping point to handle the local business, these units, together with existing dairy cooperatives, to be affiliated so as to form a central body with a small executive committee to conduct the business of the organization to the limits authorized by the producers.
6. To establish a State board empowered with the authority and responsibility of:
   (a) Enforcing the health and police power of the State as applied to milk, no milk nor cream to be permitted to enter the State unless subjected to the same regulations by New York State inspectors as are imposed on our own producers.
   (b) Deciding disputes concerning price whenever challenged, the law to provide, however, that no modification of the price shall be permitted unless the challenger shows by competent evidence that the said price would give dairymen a higher reward for labor, management and invested capital than the average awards to other industries in the State."

This group was the nucleus of the Milk Committee. Members of the Republican campaign committee approved the statement but some of them wanted to amplify it as a political expediency in a way to make it meaningless. Seeing that it was going to embarrass Chairman Eaton, personally a sincere friend of the dairymen of the State, I asked the privilege of withdrawing the proposal. Shortly thereafter the Milk Committee sent a reprint of the above statement of principles to 20 per cent of the dairymen in 22 dairy counties of the State along with a ballot containing this question:

"Are you in favor of a law to empower farmers to determine the price to be paid for their own milk as outlined in the reprint above. Check your preference."
     ☐ Yes.   ☐ No.

The distribution of these ballots was made to every post office in these counties in proportion to the number of dairymen at each post office. The replies were 99.6 per cent "Yes."

During the summer of 1936 the Milk Committee organized a working group in each of the major dairy counties, the membership was increased to nearly 20,000, a charter was prepared embracing the principles already approved and candidates for members of the Legislature were interviewed. The dairy farmers pledged themselves to vote only for candidates who promised to help perfect the bill without violating its basic principles and to support it. Fifty-two of the candidates readily promised support. Their names and many of their pictures were printed in *The Rural New-Yorker* reaching substantially all the farm homes of the State.

The Milk Charter Bill.

The bill was known as the "Milk Charter" because it provided for a charter for the dairy industry of the State, just as other industries have been provided with charters suited to their peculiar needs. The Milk Charter incorporated a local association of dairy farmers at every milk receiving plant in the State; and one central corporation affiliating all the locals in one body to perform such service as the local corporations directed it to do. The ballot vote of the local producers would control and direct the affairs of the local corporation, each one of which would be an independent unit in itself. The ballot vote of these local producers would select a delegate to organize and direct the central corporation. Provisions were made that the producers would have a direct voice in the price and terms for the sale of their milk; and to fix the expense of the whole organization. Every member was to have full information and a full detailed accounting in a simple monthly profit and loss statement.

This charter would incorporate a system for the control of the marketing and sale of milk and its products by the producers and owners of the products. It would cost nothing but the filing of a certificate for each corporation at no expense but the filling out of prepared blanks by the officers of the corporations. It required no financial wizard or business expert to realize the importance and value of this system to the dairymen of the New York milk shed.

Invitations were issued, publicly and privately and many times repeated, for criticisms of the Charter and for suggestions for changes in it or for an entirely new bill if something better could be produced and yet preserve the basic principles already approved by dairymen. These invitations were made directly to dairy farmers, dairy leaders and executives.

The bill was introduced in the 1937 Legislature by Senator Francis McElroy, of Syracuse, and Assemblyman Fred Young, of Lowville. It then became the McElroy-Young Bill.

Strategy of the Opposition.

For sixteen years the Dairymen's League had no organized legislative opposition. Its officers devised the Co-operative Corporations Law and revised and amended the Agriculture and Markets Law from time to time to suit themselves. Every new law and revision strengthened official and dealer control more than before and restricted or annulled the powers that farmers previously enjoyed. But the sudden interest of legislators in the Charter Bill and its popularity in farm circles threw consternation into the ranks of the big dealers and their allies. Their immediate problem was to gain time.

Governor Lehman said he would give farmers what they wanted. A legislative committee was appointed to find out, as alleged, what farmers wanted. Some of the most ardent proponents of the Borden-League combination were on the committee. Not a single outspoken proponent of the Milk Committee or of its Charter Bill or of the large independent groups was on the committee. The atmosphere of the hearings was created by spokesmen of dealers and dealer-controlled co-operatives. The testimony revealed a common source. Men who went to the hearings with a purpose to speak out the truth were deterred and silenced by the scope and rules of the committee. The carefully prepared testimony formed the bulk of the discussion. It was largely focused against the Milk Control Law.

The dealers and their country allies did not propose a substitute for State Milk Control, which they opposed. Neither did the Legislative Committee suggest any system of farm control to take the place of State Control, nor did they advise continuance of State milk advertising at farmers' expense.

The big dealers and their country allies soon discovered the merit and the popularity of the Charter bill. They just had to do something which had the appearance of helping the milk producers, but which at the same time left control safely in dealers' own hands. After many conferences at Syracuse and elsewhere they met at the State College in Ithaca, and, after collaborating with their College allies, put the last touches on it. It was carried to Albany by one of the College group and was first referred to as the "Ithaca-Cornell Bill." It was sent to Senator Rogers for introduction. He stated in the Senate that it was not his bill. His first correct reaction was that it "was no good." The heat was turned on later.

The proponents of the Cornell Bill announced that it was approved by the State Grange, the Farm Bureau, the Fruit Growers' Association, the G. L. F. Exchange and the Vegetable Growers' Association, the Dairymen's League and the Sheffield Producers' Co-operative Association. This was not true.

The heads of these associations together have what they call a farm conference board. The members of this official group did approve the bill, but it was never submitted to the members of these bodies for debate or approval. The heads of these organizations for the time being had a perfect right to approve the bill individually, but no one of them had authority to speak in this matter for the members of his organization. Some of these leaders had personal interests at stake, some had no free choice; their profitable jobs depended on it and the approval by others seemed to be mere courtesies to their associates in the official council. Therefore, it was a deliberate misrepresentation to say that the bill was endorsed by farm organizations.

League producers and Sheffield producers, who favored the Charter bill, protested that they were practically ordered to vote approval of the dealers' proposals at staged local meetings. In one location the producers appealed to a sympathetic friend who was not a dairyman to appear at a meeting and protest on behalf of producers, who felt it would jeopardize their interests to make their own protests. This information was sent to me in confidence with the name and address of the friendly spokesman, who, of course, could register no effective protest for an unidentified group of intimidated voters fearing threatened reprisal. The meeting reported unanimous approval of the members. That might mean a half dozen ayes and no nays.

Governor Vetoes the Charter Bill.

The Dairymen's League officials led the fight against the Charter Bill and for the Ithaca-Cornell Bill. Its lobbyists were everywhere and Albany afforded no form of pastime or entertainment that these lobbyists failed to patronize as hosts to helpful people. It was the first time the League officials were challenged to a real contest and they used milk money freely. A resolution by Senator Graves to investigate the "slush fund" was pigeon-holed in a committee, but at the same time precipitated many Albany hotel check-outs by the milk lobbyists.

The original bill gave producer and dealer bargaining agencies authority to fix the price to be charged consumers. The city consumers' associations and city legislators objected. The bill faced defeat. For two days and a night its promoters were in a dither, but the objectors stood firm and the provision was stricken out just before the vote. Farm-elected legislators may yet gain courage to protest legislation detrimental to farmers.

The bill authorized classification and blended prices and fostered monopoly. It deprived the dairy farmer of any voice in determining the price for his milk, and it sanctioned the privilege of both elected and self-perpetuating leaders to refuse an accounting of dairy farmers' milk and money. The opposition lost in the Legislature in the fight on the Charter Bill. The Legislature passed both bills and sent them to Governor Lehman.

It was promised by legislators, who assumed to speak for the Governor, that he would hold a hearing before acting on either bill. He signed the Cornell Bill, which had in the meantime become the Rogers-Allen Bill. He vetoed the Charter Bill.

It should be noted that aside from political considerations it is public information that Governor Lehman is a member of a large investment banking house, Lehman Brothers, which helped finance the National Dairy Products Corporation in the amount of $35,000,000, and his nephew has been a director in that corporation from the first. The subsidiaries of National Dairy strenuously opposed the farm Charter Bill.

CHAPTER XXVIII

THE ROGERS-ALLEN LAW

The Bargaining Agencies.

The Milk Control Law expired on March 31, 1937. The Rogers-Allen Law became effective on May 19, 1937. It is still the law. It authorized the division of the State into marketing areas and the organization of a dealers' bargaining agency and a producers' bargaining agency in each area. Any distributor in the area may be a member of the dealers' bargaining agency. Each member is entitled to one proxy vote for every 10,000,000 pounds of milk handled. A fractional vote is allowed in case the dealer handles less than 10,000,000 pounds. Only corporations organized under the State co-operative laws or similar statutes of other states may become members of a producers' bargaining agency. Each co-operative has one vote for every one hundred producer members in the co-operative. If less than 100 members, a fractional vote in proportion to such producers is allowed. Membership in both bargaining agencies is voluntary.

The law exempts dealers from the provisions of the Donnelly Act relating to monopolies, combinations or agreements in restraint of trade. Producer organizations were exempted by a previous law, but they were also included in this law.

The Rogers-Allen Law anticipated the failure of the bargaining agency set-up in providing for an order fixing the price for milk by the Commissioner of Agriculture upon the petition of a producers' bargaining agency. The Commissioner must call a hearing and if he is satisfied that the conditions in an area are as the petition alleges and 75 per cent of the producers approve, he may issue an order fixing minimum prices to be paid to producers.

Legalized Robbery.

A further alternative is provided for in the event of a Federal-

State Order fixing the price of interstate milk. If approved by seventy-five per cent of the producers affected, the Order can provide for an equalization of prices of "all producers of the production area of the market affected so that each producer or co-operative association shall receive the same base price for all milk delivered subject to reasonable differentials for quality and location and for services." After the Commissioner has fixed a price in any area or approved a price in the marketing agreement to be charged or paid for milk in any class, grade or use, the law provides that it becomes unlawful for any dealer to offer less or to pay less than the price fixed by the agreement or order.

Subdivision 9 of Section 258-m is a gem. It provides that the marketing agreement or the order of the Commissioner may provide for deductions from payments to producers for marketing service, and that no marketing agreement or order shall prevent a co-operative association from blending all its sales and distributing to its producers the blended price subjected to deductions and differentials as provided by its contracts with its producers.

All of this detail of contracts, agreements, hearings, orders and court references was intended to bewilder and confuse everyone but the authors of the law. It was a plausible attempt to appear fair to all concerned, but every provision of it strengthened the power of the racketeers, the self-appointed co-operative leaders, and the milk monopoly. As a whole, it stripped the dairy farmer of any control whatever over his milk, the price of it, or the returns for it. Instead of the same price for all producers, by the device of differentials, deductions, classifications, blendings, and special rulings a different price can be made for each producer or class of producers for each location and each plant. Conscious of its violations of elementary law, an attempt was made to overcome these defects by inducing dairy farmers to sign agreements authorizing these discriminations and a surrender of the farmers' legal and moral rights.

The proponents of the Rogers-Allen law knew that the Bargaining Agencies could not be successful. Their purpose was to defeat the farmers' Charter Bill, which would correct the faults and abuses in the existing law and would give existing co-operatives, as well as newly organized co-operatives, an opportunity for economic and efficient management under the direction of farmers themselves. The provision to bring the Commissioner of Agriculture into the picture again was adopted for the double purpose of making it possible to pass the bill in the Legislature and to lay the foundation for a new plan when the Bargaining Agency scheme failed.

The Federal Government had tried to get the consent for regimentation of dairy farmers under the A.A.A. in 1934. Farmers would have none of it. Practically the same thing was attempted in the Seven-States Pact, which the New York dealers and subservient leaders had induced the Government to foster in 1936 and which farmers turned down with a second bang. But persistence was necessary to save the dealer-leader milk monopoly and so a provision in the 1937 Federal A.A.A. law authorized a joint Federal and State agreement for the control of interstate milk. This was made to dovetail with the authority vested in the New York Commissioner of Agriculture by the Rogers-Allen Law.

Peter G. Ten Eyck's Observation.

The basic difference between the Charter Bill and the Rogers-Allen Law has been the bone of contention between dairy farmers and milk dealers for 65 years. The Charter Bill represents milk-producers' determination to recover their inherent right to negotiate the price and terms for the sale of their milk; it thwarts the crafty and cryptic schemes adopted by milk dealers to deprive the dairy farmers of the fruits of their enterprise and labor.

The pretense of the proponents of the Rogers-Allen Law is that farmers control the co-operatives and therefore have an equal voice with the dealer in negotiating the price of milk. If that were true, there would be no objection to their bill, but then neither of these bills would be needed. The "joker" is that the big dealers have power to fix the price for the co-operatives that supply their milk and this virtually fixes the price for all co-operatives and all producers. Mr. Peter G. Ten Eyck, then Commissioner of Agriculture, who had their sworn records in his office for two years, made this public statement:

"I make no distinction in the action of those who represented the dealers or those who represented the co-operatives in this unholy alliance which they formed to work out their own selfish individual ends. . . . Each of those leaders of the dealers and co-operatives is still angling for a position of vantage . . . we find that the dealers practically control all producers, producers organization and all the co-operatives. . . . They have used arguments and coercion, but principally threats to choke this belief (that farmers would have a voice in fixing the price) down the farmers' throats."

Neither Senator Rogers nor Assemblyman Allen, who sponsored the Ithaca-Cornell Bill, attempted to defend the bill on the ground that it gave farmers power to negotiate the sale of their milk. They merely insisted that they believed the system of bargaining proposed would secure a fair price for the producers. It did not and it never will, but if it had, it would still deny the farmer his inherent right of economic freedom, which is essential to personal liberty.

The Big-3 Domination.

The Metropolitan Co-operative Milk Producers' Bargaining Agency was organized in June, 1937, as authorized by the Rogers-Allen Law. It reported that Borden's and the Dairymen's League represented 24,000 producers and that Sheffield Farms represented 13,000 producers. That left about 30,000 producers in the area unorganized or in small associations.

The smaller co-operatives and the unorganized group held out because the Big-3 would dominate the Agency. The Rogers-Allen Law was a failure at the start without them. Hence a compromise was made. The Big-3 waived their domination in part, at least for the time being. By-laws were adopted which required two affirmative actions to increase or decrease the price of milk. In one action each co-operative has one vote and the privilege of withdrawing on thirty days' notice. In the other action each co-operative has one vote for each 100 producers under contract with the co-operative. In both actions a proposal to increase or decrease the price of milk is defeated when it receives less than 75 per cent of the votes in one or both of the actions.

The change indicated the suspicion of the small co-operatives and the necessity of the Big-3 to placate them for organization purposes, but the change still left the Big-3 in the driver's seat. With the alleged concessions a considerable number of the co-operatives joined the Agency. The concession was clearly an acknowledgment that under the Rogers-Allen Law the League and Sheffield did have the power of control in the Producers' Bargaining Agency.

As a matter of actual fact, however, this concession was nothing more than a joker since the Big-3 still retained their veto power and, if they so desired, could amend the by-laws under the powers given to them by the provisions of the Rogers-Allen Law. The control, therefore, never left the Big-3 and the record shows that whenever the occasion required, they have exercised that control to the detriment of the small co-operatives whom they had inveigled into membership on the strength of the joker concession.

At a dairy meeting in Lafargeville, N. Y., Justice Edward N. Smith, one of the patrons of the local plant, disposed of the voting concessions in the Bargaining Agency with a few words. He said:

"There are two co-operatives that control the whole situation (Sheffield and the Dairymen's League). The statement to us that the two great co-operatives have given up their voting rights on the directorate in this agency is ridiculous. Those who control the most milk will control the organization."

In the report of his investigation, dated March 8, 1938, Attorney General Bennett disposed of this pretense and sham of the Bargaining Agency fully and completely. He called attention to a reduction in the price of milk to farmers in December, 1937, from $2.63 per hundred pounds to $2.40, and another reduction on January 15, 1938, to $2.20. He says:

"Upon investigation into the reason for the approval of these two decreases in price, a unique situation was found. In ordinary business transactions, it would be supposed that when a price agreement has been reached, it would continue for a fixed period of time or until a new agreement had been reached. However, this is not the case with respect to prices agreed on by the Bargaining Agency. An agreed price continues only until either party to an agreement requests that new negotiations be entered into. If after such request for opening of negotiations no agreement is reached, the effect is not to leave the old agreement in status quo, but results temporarily in a breakdown of the price structure of the Bargaining Agencies. The understanding is that after such a breakdown, co-operatives are unable to bargain collectively with the distributors but must make individual agreements for price. A comparable situation would be a union contract which provided that in the event of a request for change by either party and the failure of agreement after negotiations, that the union would lose recognition and that each member would have to make his own wage agreement."

The Report then says that the very purpose of the Bargaining Agency was to give farmers a chance to secure a better price, but under such conditions the distributor is in the "driver's seat," and the producer has no choice but to take what the distributor wishes him to take. Most dairy farmers knew in advance that the results, as shown above, are just what the chief proponents of the Rogers-Allen Law intended them to be.

The Report continues:

"Witnesses testified that the proposal for a price change from $2.63 to $2.40 was defeated decisively at a delegate meeting of the Metropolitan Milk Producers Bargaining Agency. These witnesses said that after that vote one farmer representative rose and pointed out to the delegate body that the failure to agree upon a price meant that each co-operative had to go out and make peace on the best terms that the distributor would care to give. Then . . . upon reconsideration, the price reduction was passed unanimously. Under these conditions, the producer can scarcely be said to have any voice in the price he is to receive for his milk."

The Report further says:

"The charge of dealer domination has been made against the Metropolitan Milk Producers' Bargaining Agency. There can be no question but that both the Dairymen's League and Sheffield Producers' Co-operative Association, Inc., if and when acting in concert, can control the Metropolitan Milk Producers' Bargaining Agency. However, at the present time, the question of whether there is planned dealer domination of the Metropolitan Milk Producers' Bargaining Agency is academic. Actual domination has resulted under the price system."

Description by Attorney General Bennett.

The distributors in the Metropolitan area formed the Metropolitan Milk Distributors' Bargaining Agency for the purpose of bargaining with the producers pursuant to the Rogers-Allen law. It had forty-odd members at the outset. The membership is limited by law to milk dealers who deliver milk to stores or consumers in the Metropolitan area from a milk depot or a milk plant owned or operated by such a dealer. The members of this bargaining agency sell in excess of 85 per cent of all milk sold in the Metropolitan area. The one vote for each 10,000,000 pounds of milk or milk equivalent handled by a dealer is called the "volume vote."

No action with respect to marketing agreements, federal or state legislation or questions of policy relating to marketing agreements can be taken by the Distributors' Bargaining Agency unless such action is approved by 75 per cent of the "volume vote," and, in addition, 51 per cent of the members whose vote is counted numerically. The combined "volume vote" of Sheffield and Borden's is in excess of $66\frac{2}{3}$ per cent. Each one of the Borden or Sheffield subsidiaries or affiliates doing business in the Metropolitan area is considered as a separate member. Naturally, the purpose of the Metropolitan Milk Distributors' Bargaining Agency is to co-operate in efforts to buy milk from the farmer at a price most advantageous to the members.

The Dairymen's League with its subsidiaries is the third largest distributor in this district; but it did not avail itself of the privilege of membership in the Distributors' Bargaining Agency. If it had and wished to do so, it might have been a factor in increasing or decreasing the price to be paid producers.

A Fixed Principle.

Shortly after the Producers' Bargaining Agency began to function, I received a letter from Charles H. Baldwin, its executive secretary, requesting me to help organize local units of dairymen as proposed by the Charter Bill, and affiliate them with the Agency. For fifteen years I had publicly and privately refused to help selfish leaders, in alliance with dealers, to exploit dairymen, but repeatedly promised to support to the limit of my powers any dairy organization controlled by dairy farmers themselves. In the following reply to Mr. Baldwin I tried to give the proponents of the Rogers-Allen Law an opportunity to re-unite New York dairymen if they wished to do so:

July 12, 1937.

Mr. Chas. H. Baldwin,
Onondaga Hotel,
Syracuse, N. Y.

Dear Mr. Baldwin:

I thank you for the information included with your good letter of July 7th. I shall be glad to have the further information as and when available.

With reference to your request for me to support you in the work you are doing and to urge all producers to form plant co-operatives and join the Bargaining Agency, I assure you that nothing would please me more than to be able to do both of these things without violence to my own convictions and without peril to dairy farmers. There should be a way to make it possible for me to do so. Taking the purposes and principles professed by you and the proponents of the Rogers-Allen Law at face value they are the same as mine, and the same as held by a major number of New York dairymen. But I cannot harmonize what your group professes with what they do. Perhaps a frank statement of a few of the major discrepancies may lead to a clarification or modification of the differences which now disturb many dairymen as well as myself.

On information and belief I am convinced:

1. That the Rogers-Allen Bill was devised as a spurious imitation of the Charter Bill and to defeat the demands of farmers for the right to determine the price for their milk; and, while posed as a producers' bill, the real purpose of it is to gain a new lease of power for the dealers who control co-operatives and fix milk prices to both producers and consumers.
2. That price exemption, classification and blended prices are abhorrent to co-operation, that they defeated State Control and that they are the most effective means of dealer control of prices ever adopted.
3. That the Borden-League combination deliberately broke up and destroyed our once united organization. That the big profits to Borden's, and the millions absorbed by the League management is the cohesive force of that alliance, and that dairy farmers cannot prosper while it dominates the dairy industry.

4. That, on the theory that farmers would control the organization, the State enacted laws giving co-operatives extraordinary powers, but the officials usurped the powers of the members and the farmers have no control of prices or terms and get no information to fit them for intelligent control of their business in any event.
5. That your whole system is co-operative only in name, that it does not safeguard the interests, rights or welfare of the farmer in any particular, but it does provide everything needed by the management and the corporation to exploit the producers, and that through this set-up and multiple voting by delegates, enough dealers have gained power over the corporations to fix the price under your Bargaining Agency.
6. That in short your elaborate set-up is a very comfortable sinecure for a large group of leaders and their satellites, and a rich source of power and profit for big milk dealers, but it is an institution of helpless despair to dairy farmers.
7. That as now set up, your Agency affords no local co-operative and indeed no producer in or out of co-operatives, any real service for the dues he is required to pay. They are not protected in market, in prices or in terms. They are always a minority. If they help the big fry drive out the small dealers, their last state will be worse than the first.

You and your principals have the records. Show me by evidence satisfactory to an impartial mind that I am in error and I will do all that you ask.

Or, if you hesitate to reveal the record: 1. Break all alliances with milk dealers, 2. organize plant co-operatives, 3. put them and the whole system in control of members by ballot vote, 4. give every member access to full information that his vote may be intelligent, efficient and economic, 5. give dairymen full and accurate monthly accountings of their milk and money.

In other words, give us real farm co-operation and I will support you to the limit of my ability and my resources.

Very truly yours,

JOHN J. DILLON.

No reply has ever been made to that letter.

CHAPTER XXIX

BARGAINING AGENCIES COLLAPSE

Masters of the Industry.

With their Rogers-Allen Law in effect and their controlled bargaining agencies set up, the same dealers with their subservient co-operative-leaders, who had dominated the dairy industry since 1921, were now masters of the industry. They were free from whatever little restraint that State Control had held over them. They had state laws devised by themselves and contracts with producers written by themselves. They controlled the major part of the country plants. They controlled most of the co-operatives. They had classification, sanctified by law, giving them power to fix a price to the farmer to suit themselves. They controlled the centralized co-operatives. They were in position to influence most of the local co-operatives because a large number of independent dealers had joined the Distributors' Bargaining Agency.

The Metropolitan area consumes substantially 3,000,000,000 pounds of fluid milk, exclusive of cream, ice cream and other milk products. Borden's and Sheffield Farms handle 76 per cent of the retail delivery to homes. The Big-3 handle substantially all the "advertised" brands. These brands were invented by the Control Law to authorize the Big-3 to charge consumers one cent a quart more for the same quality of milk that other dealers sold for a cent less. There is practically no competition on the retail fluid milk and "advertised brands." The Dairymen's League and its subsidiaries, as a dealer, sold "advertised" and "unadvertised" brands. Borden's and Sheffield Farms sold 58 per cent of the total fluid milk distributed in the Metropolitan area.

Farmers Gypped Again.

In April, May and June, 1937, the Big-3 had made two reduc-

BARGAINING AGENCIES COLLAPSE

tions in consumer milk price in the Metropolitan area and the same number of cuts to farmers supplying that area. Subsidiaries of the Big-3 had cut prices in the wholesale store trade in bottles and bulk to make an excuse for the low price to farmers for all the milk. The purpose of this was:

1. To increase the Big-3's profits.
2. To make the first bargaining agencies' price in July seem good in comparison to the cut-throat prices for April, May and June, and
3. To lay ground for the argument that farmers should accept the bargaining agency scheme because conditions could not be any worse and might be better.

The bargaining agencies fixed the Class 1 price for July, 1937 at $2.00 per cwt. The average blended price, as announced on August 25th by five of the largest dealers, was $1.61. This price pleased no one. Farmers who had hoped to set their own price for milk were disgusted. In the meantime Sheffield Farms notified over two thousand producers in its twelve plants in St. Lawrence, Franklin and Clinton Counties that no milk would be shipped from these plants after August 1st. Their membership in the Sheffield Producers' Association was cancelled. These farmers were fighting mad. They began to join local units of the Dairy Farmers' Union. Other dairy farmers sympathized with them and extended the organization over a considerable area of the State. A strike was waged against the twelve Sheffield plants which had turned to making by-products, but got little or no milk. In some places farmers have built new plants of their own. The protests reached the proportions of a rebellion. The public began to realize that farmers had been gypped again.

Under all this pressure the Class 1 price was increased in September up to $2.35 and in November to $2.63, which was 27 cents under September-December of the year before. But the price was not satisfactory to the Big-3 who had created the bargaining agencies for their own benefit. On January 1, 1938 the price dropped to $2.40, and from January 15th to February 28th to $2.20. Attorney General Bennett shows how these reductions were made under the dealer domination of the bargaining agencies. The producers had a choice of accepting the dealers' price collectively or of accepting any price the dealers wished to pay farmers individually. They could hang themselves together or be hung separately by the dealers.

The comparison of Class 1 prices per hundred pounds for the nine months of the Bargaining Agency régime with the same months for the previous year under State Control, and also the Dairymen's League net cash basic price to producers for the same months and years follows:

Month	1937 Class 1	1937 D. League net cash	1936 Class 1	1936 D. League net cash
July	$2.00	$1.34	$2.45	$1.50
August	2.00	1.50	2.45	1.75
September	2.35	1.64	2.70	1.68
October	2.35	1.745	2.90	1.63
November	2.63	2.085	2.90	1.67
December	2.63	2.065	2.90	1.67

	1938		1937	
January	$2.30	$1.825	$2.90	$1.61
February	2.20	1.715	2.90	1.61
March	2.20	1.465	2.90	1.45
9 months' averages	2.295	1.708	2.777	1.617

According to Attorney General Bennett's Report the net basic average price paid producers for the year 1937 by the Dairymen's League was $1.554 and by Sheffield Farms was $1.912. The Sheffield price was paid in all cash. The Dairymen's League paid cash, less a deduction for capital purposes, for which, at the end of the year, certificates were issued to run for ten years in a revolving fund. During this year these deductions ran from 5 to 7 cents per hundredweight.

Borden's and Sheffield fixed the retail price of Grade B milk for 1937 from March 31st to December 31st at an average retail price of 13.6 cents a quart. The profit was 1½ cents a quart or 70.5 cents a cwt., equal to 13 per cent. This was after allowing their own figures for the cost of distribution.

Price Juggling.

The record of this milk price juggling from April 1, 1937, the first year of the Rogers-Allen Law, to April 1, 1938 reveals the monopoly's purpose under the Rogers-Allen Law and the Bargaining Agency set-up. The following data is taken mainly from Attorney General Bennett's Report:

	Milk Dealers' Gain cwt.	Loss cwt.	Price to Producers cwt.	Price to Consumers quart
On April 1, 1937, price to producer reduced 85¢; price to consumers reduced 47¢	$0.380	.....	$2.05	12¢
On April 18, price to consumers reduced 47¢; on May 1, price reduced to producers 40¢	.....	$0.070	1.65	11¢
On July 1, price to producers raised 35¢; on August 1, price to consumers raised 47¢	0.120	.....	2.00	12¢
On August 25, price to consumers raised 47¢; on September 1, price to producers raised 35¢	0.120	.....	2.35	13¢
On November 1, price to producers raised 28¢; price to consumers raised 47¢	0.190	.....	2.63	14¢
On January 1, 1938 price to producers reduced 23¢; price to consumers reduced 23.5¢	.....	.005	2.40	13½¢
On January 15, price to producers reduced 20¢; price to consumers reduced 23.5¢	.....	.035	2.20	13¢
On March 28, price to producers reduced 20¢; price to consumers reduced 23.5¢	.....	.035	2.00	12½¢

Net gain to the dealers for the year was 66.5 cents per cwt. (1½ cents per quart); net gain to the consumer 23.5 cents per cwt.; and net loss to the producer 90 cents per cwt.

The farm revolt against the Bargaining Agency's prices and a reduced fall supply forced the Bargaining Agency to increase prices for the months of November and December, 1937. Their price objective won, the farmers' strike ended. Promptly the dealers grabbed the control lever of the Bargaining Agency machinery and forced ruinous prices on producers for the first eight months in 1938.

Mr. Fred Sexauer, president of the Dairymen's League, persisted, after the first price cuts, in claiming that the Rogers-Allen Law was a great success. On January 25, 1938, commenting on November and December prices in a letter to members, he admitted that during his leadership, farmers' prices had been so low that farm buildings depreciated, farm tools had worn out and were not replaced, family health was neglected and farm children were denied their rights; but now, he alleged, prices were at a reasonable level. "This did not just happen," he said. It was "experience and knowledge and information" that created the wonders in milk marketing.

But after having voted to reduce the January price, Mr. Sexauer cautioned that dealers wanted to destroy farmers' bargaining power; they were trying, he said, to destroy the structure. The Bargaining Agency had put prices up but the Dairy Farmers' Union, which was organized in protest against low prices, had, according to Mr. Sexauer, broken the prices in the market that he had "bought and fenced in."

Sheffield Expelled Members.

It was after affiliating with the Bargaining Agency that the Sheffield Farms Company had cut the patrons of twelve plants out of the New York market and proposed to pay them manufacturing prices. This is significant in view of Mr. Sexauer's complaint against the Dairy Farmers' Union. Mr. Sexauer continued his monopoly ambitions after his fix-up of the Bargaining Agency scheme. This was the policy of the Borden-League alliance from its first inception in 1922. He then told farmers he had built a fence around the New York market, that it was the "best market a bunch of farmers ever owned," that no outsider could get in and that the groans of the non-poolers could be "heard all over the State." That was his conception of dairy farm co-operation. All must submit to his trickery, deception and exploitation or go out of the milk business. It is the concept that destroyed the farm unity of New York dairymen. It will destroy unity in any industry.

The New York farmers did not lie down in 1922 and permit the Borden-League octopus to run over them. Normally in such cases the discarded members would force a competition in the market but in this case Borden's had the power to fix their own price and they fixed it low. The League's extravagance and waste took a big toll to come out of Borden's low price and this made it possible for independent dealers to pay their producers month after month without exception for 16 years a higher price than the League producers received.

The Bargaining Agencies broke down in the early months of 1938. They announced no prices between February and September. Borden's and Sheffield fixed the retail prices as before and since. Prices to producers were deliberately, intentionally and ruthlessly slaughtered during the Summer to pave the way for the Federal-State régime in September, 1938, just as had been the case the year before to promote the Bargaining Agency régime in 1937. During both of these periods dealers' spread and dealers' profits increased.

For a time League producers accepted excuses and promises, but later their protests and withdrawals gave Mr. Sexauer some anxiety. Borden's rebates had to be paid. Minority control and political propaganda cost money, so a way had to be found to compel all producers of the State to help him pay his outrageous expenses plus Borden's rebates. The answer to that necessity was the equalization scheme and the other preferential payments to the League authorized by the Federal-State Orders.

CHAPTER XXX

THE FEDERAL-STATE ORDERS

Farmers Denied Vote.

The Dairymen's League officials and other proponents of the Rogers-Allen Law, the Federal Secretary of Agriculture and the New York State administration had failed previously in two separate attempts to gain authority to regiment New York dairymen. They laid the groundwork for a third attempt in reciprocal provisions in the Federal Agricultural Marketing Agreement Act and in the Rogers-Allen Law. These twin legal provisions were not debated either in Washington or in Albany. It was a complicated alternative to the Bargaining Agencies and no one bothered about it. But when the Bargaining Agencies collapsed, an application was made for the joint Federal-State Orders.

The Federal Order was fixed up by Allen D. Miller, a Dairymen's League attorney. It is a lengthy complicated document. I have never known anyone that claimed he fully understood it. That includes business men, lawyers and judges as well as farmers and milk dealers. Previously rehearsed hearings were held on it. It was revised. It too, like the original Rogers-Allen bill, contained authority for the fixing of retail prices to consumers in the New York Metropolitan area. It was very cleverly disguised but the purpose was detected. Dr. Caroline Whitney, of the Milk Consumers' Protective Committee, protested and it went out, but yet the revision as a whole was even worse than the original.

Authorities in Washington and in Albany and proponents of the scheme used official pamphlets, newspaper propaganda, speeches and radio broadcasts throughout the State to convince farmers and the public that the Federal Order could be effective only after being approved by two-thirds of the producers qualified to ship milk to the Metropolitan area from the seven states of the milkshed. Nor even then unless its counterpart, the State Order, was approved by three-fourths of the producers in New York State qualified to ship to the same market. In New York State many farmers said that it was useless to waste time and expense for voting because farmers were so much opposed to them there was no danger of the Orders being approved.

The Federal and State authorities urged persistently that this plan was something for the farmer; something, they said, to solve the hoary milk problem. Farmers alone were to get the benefit and there were no farmers' responsibilities. Farmers were led to expect $2.25 per cwt. for all milk and $2.40 was suggested as possible. The cost was to be paid by dealers. There would be no restrictions whatever on farmers. The purpose was to restrict and regulate dealers. Prices to all farmers would be uniform.

Mr. Holton V. Noyes, Commissioner of Agriculture, proposed that every farmer cast his own vote. He declared that unless this were done, he could never know whether farmers really wanted the scheme or not. Later he consented to the multiple vote by the co-operatives, and exercised his discretion to announce a favorable vote on the strained assumption that if all producers had voted in proportion to the mass vote cast by the co-operatives, the vote would be overwhelming in favor of Federal-State Control.

The Dairymen's League officials and some other co-operatives insisted that the law authorized the co-operatives to vote their memberships in bulk. They demanded the right to exercise that privilege. It seemed strange to some of us with long dairy memories that the dealers, who were said to be regulated so much by the Orders and who were required to pay so much, should be strangely silent, but their leader affiliates gave assurances that they were, of course, opposed to the agreement. Actually it was their scheme devised to head off the farmers' plan to set the price for their own milk.

Farmers Challenged the Count.

Finally the time and place were set for taking the referendum vote. The ballots were prepared. The committees to oversee the balloting were selected and tellers for the count were chosen. I have been unable to find that any farmer or organization known to be opposed to the Federal-State Orders had been consulted or considered in any way in these proceedings. The voting was conducted on August 17th, 18th and 19th. On August 26th government officials announced that the Federal and State Milk Marketing Orders had been overwhelmingly approved by dairymen and would go into effect six days later on September 1st. The count, the announcement stated, showed 33,150 in favor and 5,800 opposed. No details were given. There was no statement of the number of dairymen eligible to vote either in the State or in the Federal area. There was no promise of further details.

Dairy farmers of all classes protested. Their protests reached the press, and papers in the territory joined in the protest. More than two weeks later, details were revealed on the State vote, showing 25,935 votes in favor (19,115 voted by co-operatives in bulk). The Federal Trade Commission Report No. 95, page 8, gives the number of State farmers inspected by the New York City Department of Health, and revised to January, 1936, as 44,280. Hence the approval was 58 per cent; the vote was 17 per cent short. The law provides that the Commissioner must find that the purpose is approved by at least 75 per centum of producers in the area before he may fix and determine minimum prices to be paid the producer.

Later, a release from Washington on the Federal Order reported 33,663 "yes" votes; 24,580 voted by co-operatives and 9,083 individual votes. There were 4,964 votes recorded against the Order; 4,641 individually and 323 by co-operatives. In all, 45,644 votes were cast and 7,017 were rejected as ineligible. The Co-operative Law forbids proxy voting. One co-operative protested that it voted 395 votes against the Order while only 323 co-operative votes were reported against it.

The "yes" votes, representing 51.6 per cent of the 65,255 producers approved for the metropolitan market, were 9,405 short of the number required for approval. In the Jetter-Rock Royal case, Federal Judge Cooper found that in the August referendum 22,927, or slightly more than two-thirds of the 33,663 favorable votes, were cast by the Dairymen's League, which had only 19,500 members. He also said that "the finding that the Order was approved by at least two-thirds of the producers who voted in the referendum was unsupported by any substantial evidence."

On this state of affairs the Federal and State authorities have repeatedly proclaimed that the Orders were overwhelmingly approved by dairy farmers. No further details of the vote have ever been given. Farmers are accustomed to a report of their vote at every voting place and a summary of their counties, districts and State. They were disappointed and discouraged and fearful. The check-up of the information as revealed confirmed their worst suspicions.

Inequities and Intrigue.

The alleged purposes of the Federal-State Orders were:
1. Fix a "living price" for milk.
2. Benefit dairy farmers.
3. Increase farmers' incomes.
4. Equally share the "surplus" burden.
5. Stabilize markets.
6. Pay every farmer a uniform basic price.
7. Charge the expense to dealers.

The Orders fail in each and every one of these purposes:

(1) Farmers have not benefited nor have they received a "living price," unless that term be construed to mean a price that will just keep them alive.
(2) Farmers have received no benefits, their hardships and abuses were never greater than since the Orders were adopted, consumers' costs have increased, discouraging demand for fluid milk to the farmers' detriment, and dealers' profits have been more.
(3) Farmers' income has not relatively increased. Markets have not been stabilized. On the contrary there have never been wider fluctuations, or more confusion and disorder in the metropolitan market than since the Federal-State Orders were approved. A strike in August, 1939 shut off 60 per cent of the city's supply as a protest of a majority of the producers supplying the marketing area.
(4) The surplus is the kernel of the system. It is created and main-

tained by the proponents of the system for their own profits. It is a liability to dairy farmers. It has been a source of great profit to milk dealers, who first used it as a club to break down the price paid farmers for all milk, and to save themselves from all waste in their requirements of fluid milk and cream. Some of them have cheated the producer by reporting milk as surplus after selling it as fluid milk at Class 1 prices. They store surplus cream in flush seasons and sell it in seasons of scarcity at high season prices. They manufacture it into by-products at fabulous profits.

Farmers do not "share" the surplus burden; *they carry the whole burden.* If they were permitted to negotiate the sale of their own milk, they would sell what the dealers required as fluid milk at the cost of production plus a profit, and find other uses for what the city markets did not want at the fluid price.

(5) The market has been stabilized only to the extent that the Federal-State régime is a monopoly, barring all other producers from the area and clothing its barons with power to fix the price for both producer and consumer and to displace and limit the number of small independent dealers.

(6) The uniform price is a misnomer. It is rather a series of discriminations against producers, small dealers and small co-operatives for the benefit of the Big-3 monopoly. These discriminations include:

    (a) the iniquitous classification plan;

    (b) a 25 cents per cwt. bonus to win approval in four New York counties, in 12 townships of another county in New York State, and in 10 counties in New Jersey, Connecticut, and Massachusetts. Most of this milk sells at a premium. By taking this 25 cents per cwt. for a special class of producers out of all producers' returns, the dealers are able and free to pocket for themselves whatever extra price or trade advantage they can receive for such nearby milk in the market.

    (c) payments to co-operatives of discriminating bonuses of 23 cents a cwt. and 30 cents per cwt.

    (d) Monthly bonuses to co-operatives as such being discriminatory as against other handlers, and further discriminatory in favor of big co-operatives against small co-operatives, ranging from one cent a cwt. to 2½ cents per cwt. and 5 cents per cwt. for the favored big co-operatives.

(7) With the power to fix the price to be paid the producer in the first instance, to determine weights and fat tests, to make deduc-

tions for an unlimited number of services and to refuse an accounting, the claim that the dealer pays the expense of the complicated and extravagant system, seems like an insult to the farmer's intelligence.

Another discrimination was in the Grade A premiums. Legal milk in New York State is milk with a minimum three per cent fat content. A premium has been paid on Grade A milk because it is supposed to be handled under special regulations to keep it cleaner and cooler. But for commercial purposes the bacteria count and fat test have been juggled so that the Grade A premium has varied from 78 cents down to five cents per cwt. of milk. Some patrons have received a premium of 3.2 cents per cwt. or 7/100 of one cent per quart. The only "uniform price" here has been the three cents a quart, or $1.41 a cwt., paid dealers by all consumers. The grades were discontinued on September 1, 1940 against the protest of the proponents of the system.

According to statements made by Borden's and the Dairymen's League to producers and elsewhere by Market Administrator Harmon in the fall of 1939, Borden's seems to have paid a Grade A premium of 60 cents per cwt. for 3.7 per cent milk for which the producer got no premium, 66 cents per cwt. for 3.8 per cent milk for which the producer got 10 cents, and 72 cents per cwt. for 3.9 per cent milk for which the producer got 15 cents. None of the three statements accounted for the difference of 60 cents, 56 cents and 57 cents per cwt. in Grade A premiums. The Administrator passed this off as a matter between the Dairymen's League and its producers quite apart from his responsibility. But it does not seem to square with the alleged purposes of the system.

The so-called equalization of payments (or uniform price) is a discrimination that forces one group of dealers and their producers to pay money into a fund to be paid out of the fund to other dealers. This clearly violates the "due process" provision of the Constitution. It penalizes the competent, thrifty and industrious and rewards the incompetent, extravagant and shiftless.

The farmers' "surplus" burden is created by milk dealers with the help of dealer co-operatives. It is a source of comfort and big profit to dealers. If a farmers buys more shingles than he needs to roof his barn, he cannot pay for the surplus shingles at the price of kindling wood. The dealer wants classification because he can use all milk he can sell for fluid use and pay kindling-wood prices for what remains. He makes his biggest profits on the by-products. When farmers control their own co-operatives, they will sell the dealer his requirements of fluid milk at fluid prices, and they will make a profit on the surplus or stop producing it. Sharing the surplus burden just means to the farmer bearing the loss on classification.

Nor is all this the end of the discrimination. Handlers selling a part of their milk outside of the Metropolitan area are permitted to report such sales to the Administrator at any price they see fit and blend the price in the price of milk sold in the marketing area. This is called "unpriced milk." Only the Dairymen's League, Borden's and Sheffield Farms are able to profit materially by this privilege. The prices are higher in New Jersey and Connecticut than in the marketing area.

Suppose the New Jersey price is $3.00 to the handler and the uniform price in the New York area is $2.00. Under the privilege granted in the Order, the League need not report any of its New Jersey sales; therefore it is able to use the unreported profits on all Jersey sales to undersell smaller dealers who operate only in the New York market. Besides, the record to date indicates that no producer shipping to Borden's, Sheffield or the League has received one extra penny for his milk sold in the higher brackets outside the Metropolitan area.

The Orders allow the three big dealers to charge three cents a pound for making cheese. The regular price is one and one-half to two cents. They also have a liberal margin on the manufacture of butter and other by-products.

Nor is this all yet. These easy extra profits are used to undersell other dealers with the result that the other dealers are forced to meet the competition and take the loss out of their producers or go out of business. These conditions are now imposed by law on all producers and independent dealers whether they like it or not.

THE FEDERAL-STATE ORDERS

Judge Cooper found from evidence not disputed that the difference between the price paid by the Dairymen's League to its members for milk sold in New Jersey during September and October, 1938, and the New Jersey Control Board Class 1 price was a total of $202,447; that in addition its extra payments under the Order amounted to $258,773, which, when applied to all sales, gave the League a competitive advantage amounting to 60 cents per cwt. in September, 1938, and 47 cents per cwt in October. Yet it paid its producers 7 cents less in September and 13 cents less in October than for the same months of 1937. It paid its producers less for both months of both years, as well as for all months of all years, than other dealers paid producers.

Originally the producer delivered the milk at his nearby plant, but since local plants have been removed the producers are required to pay from six to twenty cents a cwt., according to the distance, for cartage to the centralized plant. The farmer pays this out of the price he is supposed to get. There is considerable saving in handling, processing and freight to the handler but the saving has not been reflected in the producer's price.

Every one of these discriminatory payments has come out of the farmers' milk pails. Some of the money goes to some of the favored producers, but most of it goes to pay the cost of keeping a small group in power and the monopoly in control of milk prices.

On October 25, 1938 the Administrator of the Federal-State order reported for its operations in September. The uniform price was $1.87. The farmer did not get this $1.87 in all cases. All of the milk produced by Borden's, Sheffield and League producers is "pooled" twice. Further, deductions are made by the Federal Administrator and by dealers for services and expenses from all the milk shipped to the Metropolitan area. The payment to the producer is also modified by the freight differential, the fat differential, and in the case of the Dairymen's League for certain other differentials, expenses and losses incurred as a dealer. The following dealers reported basic returns to producers for September 1938 subject to modification by the fat, freight differentials, and dealer deductions:

	Per 100 lbs.
Brescia Milk Co., Inc.	$2.14
Sheffield Producers' Co-op. Assn.	1.87
Unity Co-op. Dairymen's Assn., Inc.	1.81
Crowley Milk Co.	1.81
M. H. Renken Dairy Co.	1.80
Dairymen's League Co-op. Assn., Inc.	1.545

The proponents of this plan compared their September price with their depressed price of the August preceding and claimed a rise. But in fact it was lower than in Septembers of previous years. For the three year period the average basic price to farmers for September milk was as follows for the same dealers reporting.

1936	$2.09
1937	2.15
1938	1.82

The price announced by the Federal Administrator was 27 cents less than the 1936 price and thirty-three cents less than the 1937 price. In 1938 the consumer paid three-fourths of a cent a quart, or thirty-five cents a cwt., more than in 1937, so that for September, 1938, under the Federal-State Orders the dealers gained thirty-three cents from producers and thirty-five cents from consumers, or a total profit of sixty-eight cents per cwt.

This is the triumph of the milk racket operating within the framework of a milk monopoly legalized and abetted by a combination of Federal and State laws.

The alleged services to the industry are mere pretense. There is no good reason why expenses of handling milk for manufacture should be charged to fluid milk. Nor is there any good reason why the patrons of dealers should be penalized to pay a bounty to co-operatives. Nor again is there any good reason why a bounty of one cent per cwt. should be paid one co-operative and five cents per cwt. to another co-operative and nothing whatever to still other co-operatives. The legitimate co-operative system was devised to save money in distribution and to give farmers a collective bargaining power. If they must have a pension to survive, they fail in their purpose.

To put it all in a nutshell, bogus centralized co-operatives cannot compete with true co-operatives and dealers and cannot longer hold their producers unless they are paid as much as they get elsewhere. Hence this elaborate scheme, devised to remove their competition, creates a monopoly of the three biggest dealers, bars all farmers from the market who defy their power, actually compels one group of farmers to share their income with another group and compels all producers by transparent devices to contribute to slush funds, political needs and notorious waste and extravagance.

The Federal-State pool is a trust. The United States Secretary of Agriculture, the State Commissioner of Agriculture and Markets, and their Administrators are its trustees. None of them make any accounting of the milk or money to show the men who produce the milk what the money was spent for or who got it.

The Dairymen's League pool is a trust. Its officers are trustees. Farmers are illegally barred by contract from demanding an accounting.

These two groups of officials are the only trustees in the civilized world who are exempt from an accounting of their trusts.

These things didn't just happen by accident. The big dealers wrote their own ticket and the legislators in Washington and Albany voted it all into laws without in the least knowing what it was all about. The New Dealers on the Supreme Court admitted discrimination but did not seem to realize the gravity of it, the principles violated nor the turpitude involved.

CHAPTER XXXI

THE LEGAL ENTANGLEMENTS

The Federal Order Annulled.

Small dealers and local co-operatives paid their assessments to the Federal-State Administrator for the equalization fund on the expense assessment for one or more months. Then they began to see what would happen to them if they continued to do so. Some of them stopped the payments. A suit was filed against the Jetter Dairy Company, the Rock Royal Co-operative, the Central New York Co-operative and the Schuyler Junction Co-operative, as defendants. The plaintiffs were the United States of America, the Dairymen's League Co-operative Association, Inc., and the Metropolitan Co-operative Milk Producers' Bargaining Agency. The chief counsel for the plaintiffs were John Yost and Charles J. McCarthy, assistants to the United States Attorney General, Seward A. Miller, Edward Schoeneck, Paul J. McCauley and Edmund F. Cooke. The defense was represented by Willard R. Pratt, Charles W. Jenkins, Albert Haskell and Samuel Rubinton. The suit was tried in the United States District Court in Albany before Judge Frank Cooper. Taking of testimony occupied about four weeks, ending on January 17, 1939.

Originally the action had been instituted by the Government against the defendants to enforce compliance with the Federal Marketing Order. Because of the defendants' charges of misrepresentation in connection with the adoption of the Order, the League and the Bargaining Agency intervened to defend themselves. They, however, failed even to attempt this defense. They had been charged with conspiracy, restraint of trade and monopoly. They did not bring a single witness to deny the testimony of the witnesses making these charges. The decision of the Court was rendered on February 23, 1939. Judge Cooper held that the

Federal statute, as applied in the Federal Order for the metropolitan market, was unconstitutional and that the Order itself had not been legally adopted, and should not be enforced.

During the four weeks' trial Judge Cooper obtained a thorough grasp of all the salient facts involved. His keen penetration of these facts enabled him to render one of the most telling and sweeping analyses of our milk monopoly yet to be found in our public records. Following his decision Henry Wallace, Secretary of the United States Department of Agriculture, and Holton V. Noyes, Commissioner of Agriculture and Markets of New York State, suspended operations of both the Federal and State Orders.

The Court held that the equalization provision constituted taking property of one without compensation and giving it to another. He also held that the statute unlawfully discriminated in favor of the larger co-operatives and big dealers. He further held that there was no doubt as to the misrepresentation in the propaganda handed out by the League, the Bargaining Agency and the Federal and State Departments as to the contents and effects of the Federal Order, and that had it not been for these misrepresentations, the Order might not have been approved.

Judge Cooper's Findings.

These are some of the highlights in the Court's decision, as to motives, persons, purposes and effects:

"The Dairymen's League, in co-operation with the New York State Conference Board of Farm Organizations, actively sponsored and supported the enactment of the Rogers-Allen Law. Pursuant to the provisions of this law, the Bargaining Agency was formed. Several directors of the Dairymen's League acted as incorporators of the Bargaining Agency and were active in the creation and organization thereof.

· · · · ·

The Dairymen's League occupies a dual capacity as a co-operative agent of producers and as a distributor.

· · · · ·

It is clear that the Dairymen's League and Sheffield Producers' Association exercised an important, if not the dominating influence over the activities of the Bargaining Agency.

About November 1, 1937 the Bargaining Agency appointed a lawyers' committee to draft a form of marketing agreement and the original draft of this agreement was presented to the Committee by Allen D. Miller, one of the attorneys, for the Dairymen's League.

. . . . .

The Bargaining Agency and the Dairymen's League, in furtherance of their scheme, carried on an elaborate campaign of false and misleading propaganda through newspapers, radio speeches, letters, pamphlets and otherwise calculated to influence milk producers.

. . . . .

The Bargaining Agency, whose largest and most influential member was the Dairymen's League, spent $108,637.91 from December 1, 1937 to September 1, 1938 to promote the adoption of this Order. During the three months between June 1 and September 1, radio expenses were $7,029; and advertising and publication expense $7,979.

. . . . .

The Dairymen's League actively sponsored the proposed Orders. The *Dairymen's League News,* its official organ, actively advocated adoption of the Order.

. . . . .

There can be little doubt of the co-operation between Federal and State agents with the co-operatives. Indicative of co-operation is the misleading statement of the Department's published statement of August, 1938 taken verbatim from the *Dairymen's League News* of August 16, 1938, to the effect that the proposed Order required handlers to pay the uniform price to all producers and the operation of the pool assured payment to all producers at a uniform rate. These statements were incorrect.

. . . . .

In the referendum held August 19, 22,297 or slightly more than two-thirds of the 33,663 favorable votes, were cast by the Dairymen's League. Yet it had only 19,500 members.

. . . . .

It is reasonably clear that producers were led to believe that equal uniform prices for all producers would be received under the Federal Order.

. . . . .

We have then a favorable vote obtained or influenced by misinformation as to the terms and effects of the Order.

THE LEGAL ENTANGLEMENTS 253

Had it not been for these misrepresentations, the Order might not have had the approval of two-thirds of the votes counted as valid.

. . . . .

The finding that the Order was approved by at least two-thirds of the producers who voted in the referendum, was unsupported by any substantial evidence.

. . . . .

The finding by the Secretary of Agriculture that the Federal Order was approved or favored by two-thirds of the producers producing milk in the production area is invalid.

. . . . .

The orders were issued as a result of the program of the Dairymen's League, begun with sponsoring of the Rogers-Allen Law and continued through the formation of the Bargaining Agency, the formulation and proposal by the Bargaining Agency of the proposed Federal Order, and finally adoption of the same.

. . . . .

The purpose and effect of the Order was to give to the Dairymen's League a competitive advantage, and to that end the Dairymen's League, the Bargaining Agency and E. Manning Gaynor, personal representative of Commissioner of Agriculture Noyes, entered into a conspiracy to establish in the League, the Bargaining Agency and in dealers purchasing milk from the Bargaining Agency members, including the League, an illegal monopoly in the purchase and sale of milk."

As to price chiselling the Court found this undisputed evidence in the testimony:

"It is agreed here that dealers must receive not less than 9¾ cents per quart to make full payment under the Order to producers. Jetter Dairy Company had been selling milk to one Weintraub at 9¾ cents per quart. On September 4, 1938 Dairymen's League acquired 10 cases of Weintraub's trade by selling him milk at 9 cents a quart.

. . . . .

Sheffield Farms furnished milk to one Lavender at 9¾ cents per quart and rebated to him ¾ cents per quart on all purchases, and in addition, delivered one extra quart per case, making the net cost to Lavender 8¼ cents per quart.

. . . . .

Sheffield gave one Rose Paley 35 cases of milk free in order to induce her to buy Breakstone (Sheffield subsidiary) milk, charged her 9¾ cents per quart and rebated ¾ cents per quart. The Dairymen's League also gave Rose Paley 35 cases of free milk and charged her 9 cents per quart.

* * * *

If the Dairymen's League had to pay the same price to producers as independent dealers were required to do under the Order, it would have lost from $10,400 to $15,600 per day in making sales to stores at 1 to 1½ cents below 9¾ cents. But the Dairymen's League does not pay the uniform price to its producers and is not required by the statute or Order to do so. It may deduct such losses from the price paid to producers.

* * * *

The difference between the price paid by the Dairymen's League to its members for milk sold in New Jersey during September, 1938, and the New Jersey Control Board Class 1 price was $103,105; the difference in October, 1938, was $99,342; a total of $202,447."

Losses to League Producers.

This is the Court's record of losses to League producers for 1938 compared with 1937 when the competitive advantages on unpriced milk under the Federal Order were sixty cents per cwt. in September, 1938, and 47 cents in October, 1938, an average of $3.02 per cwt. on the volume handled as a dealer:

"The Dairymen's League gross pool return for September, 1938, was 7 cents less than September, 1937, and for October, 1938, was 13 cents less than October, 1937.

* * * *

Applied to all sales made by the Dairymen's League, its total competitive advantage from sales in New Jersey market service payments, co-operative service payments and by payments to its own producers at less than the uniform price, amounted to 60 cents per cwt. for September, 1938 and 47 cents per cwt. in October, 1938.

* * * *

Applied to sales made by the Dairymen's League as a dealer, its total competitive advantage from these extra payments amounted to $3.30 per cwt. for September, 1938 and $2.74 per cwt. in October, 1938."

The Federal and State Orders cost Sheffield producers a loss of 33 cents per cwt. for September and October, 1938, yet

"Total extra payments received by Sheffield Farms Company in September and October, 1938, under the Federal Order, exclusive of profit margin on New Jersey sales, amounted to $76,497."

. . . .

"The statement of Charles H. Baldwin, Executive Secretary of the Bargaining Agency, on November 21, that the Order had already added $2,500,000 to dairymen's incomes is contrary to the statistics in evidence."

As to unlawful benefits to manufacturers of milk; and to take property from one without compensation and give to another, the Court said:

"The large dealers, principally National Dairy subsidiaries, Borden and its subsidiaries, and Dairymen's League, as well as Queensboro Farm Products and the Aaron J. Rubenfeld Companies, all engaged in manufacturing milk, received market service payments under the Order and thus substantially reduced the prices paid by them for fluid milk.

. . . .

The Court does not find any sections of the Federal statute which clearly and definitely authorize 5 cents per cwt. payments to co-operatives and 23 cents per cwt. market service payments plus transportation charges.

. . . .

Prices fixed by the Market Administrator under the Federal Order for approved milk made into cheese were less than the prices paid by cheese factories purchasing unapproved milk to the extent of 20 cents per cwt. Borden, National Dairy and the League were able to, and did, purchase approved milk for such cheese purposes and thereby enjoyed discriminatory advantage.

. . . .

Congress has not the power under the Commerce Clause or otherwise, to take without compensation property of one and transfer it to another. There can be little doubt that the Federal statute takes the money from one group without compensation and transfers it to another."

The Order was held discriminatory:

"The (Federal) Order is discriminatory against the defendants in that moneys belonging to them are expropriated to be paid to competitors of said defendants and in effect to give such competitors, principally the Dairymen's League, Borden and Sheffield, a subsidy from the defendants' capital, whereby a monopoly may be created in said large competitors.

The Order is also discriminatory as against the defendant co-operatives (Rock Royal, Schuyler Junction and Central New York). They pay money, which would otherwise be distributed to their producers, to the settlement fund, in order that this money may be paid out to the large co-operatives and dealers.

. . . . .

The Dairymen's League, by its acts in procuring the passage of the Rogers-Allen Law, the formation of the Bargaining Agency, and through its own activities and those of the Bargaining Agency, has conspired to convert the provisions of the Federal law and of the Rogers-Allen Law as instruments of coercion, restraint of trade and monopoly. In furtherance of said conspiracy, the League and the Bargaining Agency aided in procuring the promulgation of the Federal Order.

. . . . .

The Dairymen's League and the Bargaining Agency are now engaged in an unlawful combination and conspiracy.

. . . . .

Neither the Order nor the statute made any attempt to remedy the competitive advantages and the operation of the Order must have been foreseen. In any event, the statute as applied to the Order brings arbitrary discrimination and disaster."

In the Niagara Frontier.

The Dairymen's League and the New York Department of Agriculture worked up a similar bogus producer consent for a State Marketing Order in the Buffalo, New York, area. That Order affects all the dealers and co-operatives in the area and the Niagara Frontier Milk Producers' Bargaining Agency, Inc. No interstate milk is supposed to enter this area, so the Federal government was not a party to this Order.

When an Order is issued either by the State or by the Federal and State governments jointly, the price fixed applies to all milk distributed in that marketing area, whether or not the producer or the dealer approves the price or the system. In the Buffalo area two dealers and two co-operatives refused to pay the equalization fee. Commissioner Noyes filed separate suits in the New York Supreme Court to compel each defendant to pay into the equalization fund and to restrain them permanently from violat-

THE LEGAL ENTANGLEMENTS 257

ing the Order. He also asked for a temporary injunction. This injunction was denied by Justice Francis Bergan. The joint motion of the defendants to dismiss the suit was granted on the ground that the Rogers-Allen Law, insofar as it authorized the Commissioner to fix and equalize prices under a marketing order, was unconstitutional.

The Nunan-Allen Law, enacted in 1939 by the Legislature, was devised to amend the Rogers-Allen Law in a way to comply with Judge Bergan's opinion. It changed the verbiage of the original law but in the opinion of opposition lawyers, which I share, the basic principles are the same as before. Commissioner Noyes announced that the State Order in the Buffalo area would be reinstated under the Nunan-Allen Law July 1, 1939.

Judge Bergan Reversed.

In the meantime, the New York Court of Appeals reversed Judge Bergan's decision. The Court, in an opinion by Judge Harlan Rippey, held that the Legislature has the authority under the police powers of the State to pass regulatory laws and to delegate powers to an administrative agency, in this case the Commissioner of Agriculture.

Disappointing as this decision was to dairy farmers throughout the State, the Court was careful to point out that it confined its ruling solely to "the facts set up in the complaints" and that "it would necessarily depend upon the facts and circumstances in each particular case at bar after the issues were duly presented and the facts then at issue developed whether the enforcement of the order would make a confiscation or a discrimination as to them." The door of the Court is therefore still open to producers who allege they suffer "confiscation or discrimination."

The fact that there is nothing in the Constitution specifically to forbid the enactment of a law is no proof that the law is just or fair. The Constitution is properly expressed in broad principles. Considerable discretionary authority is left to the legislature in the matter of scope and details. Legislators are expected to reflect the wishes and interests of their constituents. In this country they generally do yet reflect the audible sentiment of the people. The judiciary is one of the three co-ordinate divisions of the government. The Legislative and Executive Departments are the other divisions. While each is intended to check errors and faults of the others, the duty of each is to co-operate and not to antagonize either of the others.

The Courts are always striving to uphold a law, once it is placed on the statute books. They frankly say so; therefore, a law must be flagrant in violation of basic principles before it will be declared unconstitutional by the Courts. Another thing to remember is that the Courts must go by the evidence produced in each case. In these milk cases the theory is that the laws are all enacted for the benefit of the farmer, that the co-operatives are all being run for the best interests of the producers and that their leaders are one and all working for the farmer and no one else.

This theory goes out to the press and the public through a well-organized and well-paid propaganda machine. Its proponents put it before the Legislature. It gets into the statutes and also as testimony in the Court records. But that this theory is far from the truth is seen in a study of the dairy laws, customs and contracts now in force in New York State. That such theory is fictitious is seen in the domination of co-operatives by the milk barons. It is seen in the alliance of co-operative leaders with these magnates of the milk monopoly, and it is seen in the increased cost of the complicated machinery of distribution and the starvation prices paid producers.

The truth about the dairy industry and the hardships of the average dairy farmer have never reached the Legislature, the public or the Courts. When the facts are made known and the sentiments of these groups are focused into action, I believe that the burdens will be lifted from the shoulders of the dairy farmer and that all the people will share in his renewed prosperity.

Fraud Ignored, Indictments Stand.

In a split decision of five to four handed down June 5, 1939, the United States Supreme Court sustained the validity of the Federal Order, and to that extent reversed Judge Cooper's decision. The bare majority consisted of the four new justices that had been recently appointed by President Roosevelt, and Justice Stone. The minority was represented by Chief Justice Hughes and Associate Justices Butler, McReynolds and Roberts. The original Cooper decision invalidating the Federal Order was based principally on the fraud, misrepresentation and coercion practiced by the Dairymen's League and the Bargaining Agency and the attempt to set up an illegal monopoly with the Borden Company, National Dairy and the Dairymen's League in control.

The majority recognized the conspiracy to obtain a monopoly, yet held that such a motive was insufficient to warrant invalidation of the Order. The majority also admitted that the misrepresentation and fraud constituted a "variation from the facts" which was "not immaterial," yet held that "a study of the official forming of the Order would have cleared up any misconception created by the language." It admitted that the Order discriminated against some producers and favored others. Even though these discriminatory items ran from five to thirty cents a hundredweight, with an individual farmer possibly subject to two or more or all of these, yet the majority held that "the discrimination seemed fantastic and remote."

These judges further specifically assumed that Congress has the power to put a milk marketing order into effect without the approval of anyone, and hence argued that Congress may authorize co-operatives to cast the vote of their memberships. In like manner, notwithstanding its recent decisions to the contrary, the Court decided that Congress may delegate power to the Secretary of Agriculture to come into business, make his own laws and enforce his own decrees—to be a dictator. In short, therefore, the five men seemed to have first determined that they must reverse Judge Cooper's decision and then scouted around to find reasons or excuses for doing so.

There were two separate dissenting opinions. In the first, Justices McReynolds and Butler said that in their opinion Secretary Wallace's order must succumb to two manifest objections and that it was not necessary to dissect the record for other errors.

First. Congress possesses the powers delegated by the Constitution and no others. This view was held by the same Court in

1935 in the famous N.R.A.—Schechter poultry case which demonstrated the absence of Congressional authority to manage business affairs under the transparent guise of regulating interstate commerce.

Second. Even if Congress had power to manage a milk business, it could not commit that power to another. The statute shows clearly the design to allow a Secretary to prescribe according to his own errant will, and then to execute. This, they held, is not "government by law but by caprice" and the outcome seemed "wholly incompatible with the system under which we are supposed to live."

Justice Roberts wrote a separate dissenting opinion in which he held that the law involved an unconstitutional delegation of powers and that the Order was not authorized by the law; but that if it were authorized, it would deprive the defendants of their property in violation of the Fifth Amendment to the Constitution. He said that Judge Cooper's conclusions are based on findings of fact, which findings are based on uncontradicted testimony, authentic documentary evidence, and a stipulation of the parties, and that they should be accepted by the reviewing Court.

With respect to discriminations, Judge Roberts argued that cooperatives, such as the Dairymen's League, and also Sheffield Farms Company were favored by the Order in the sale of milk outside the marketing area. He quoted from the record of the trial from which he concluded that:

"As the Order is drawn and administered it inevitably tends to destroy the business of smaller handlers by placing them at the mercy of their larger competitors. I think no such arrangement was contemplated by the Act, but that, if it was, it operates to deny the appellees due process of law."

The majority opinion made a feeble attempt to justify this discrimination against smaller handlers by claiming that this was a competitive situation which the Order did not create and with which it did not deal. Judge Roberts' answer is irrefutable:

"The opinion of this court states that the detriment to the smaller handlers who sell milk for use only in the marketing area is the result of competitive conditions which the Order does not effect. But it is evi-

THE LEGAL ENTANGLEMENTS 261

dent that the Order freezes the minimum price which is to be paid by many handlers and leaves the price of other handlers who compete with them open to reduction by the device of blending."

The New Deal Federal Order was sustained by New Dealers who set aside time-honored American principles and doctrines, but who found it impossible to deny Judge Cooper's original findings of misrepresentation, fraud, coercion and monopoly. The testimony on which they are based was not contradicted by any witness in the trial. The indictments still stand unchallenged and undismissed.

Federal Indictments in Chicago.

In November, 1938, the Federal Government obtained 97 indictments against corporations and individuals and organizations charging violations of the Sherman Act in the distributon of milk and ice cream in the Chicago milkshed involving interstate commerce. The indictments followed a four months' investigation by Federal officers and a Federal grand jury. A year's investigation by the Department of Justice preceded the grand jury action.

There were four counts in the milk indictments. One charged a conspiracy to fix and maintain arbitrary, non-competitive prices paid to producers in the Chicago milk shed, the second charged collusion in fixing retail prices of milk; the third a conspiracy to shut independent merchants out of the market; the fourth alleged that there was a plot to limit the supply of milk coming into Chicago. New Yorkers named in the indictments include: The Borden Company; Borden Wieland, Inc.; National Dairy Products Corporation; General Ice Cream Corporation, Thomas H. McInnerney, President National Dairy Products Corporation; Vernon F. Hovey, President General Ice Cream Corporation; R. V. Jones, Vice-President Borden Company, Leland Spencer and Madison H. Lewis.

On July 13, 1939, Federal Judge Charles E. Woodward ruled that Congress had removed the marketing of farm products, including milk, from the jurisdiction of the Sherman Anti-Trust Law and had placed control in the hands of the Secretary of Agriculture. The Court held that by the Agricultural Marketing

Act, Congress has given the Executive Department, acting through the Secretary of Agriculture, full power over the production and marketing of agricultural products. The Court said:

> "The marketing of agricultural products, including milk, covered by the agricultural marketing agreement act, is removed from the purview of the Sherman Act."

and that

> "The Secretary of Agriculture was empowered to control the industry in any milk shed and it was his duty to intervene if the act was violated."

The Attorney General appealed and the Supreme Court in December, 1939 in an unanimous decision reversed Judge Woodward and reinstated these Chicago milk monopoly indictments. The Court held that the co-operative-dealer-union combine could be prosecuted under the monopoly statutes. For reasons stated, the Supreme Court held it could not be said

> "that commerce in agricultural commodities is stripped of the safeguards set up by the anti-trust act and is left open to the restraints, however unreasonable, which conspiring producers, distributors and their allies may see fit to impose."

Suddenly and without explanation, an announcement was made in the public press in September, 1940, that this prosecution had been settled with the defendants signing a "consent decree" said to be satisfactory to the Federal government. The published reports indicated that under the decree, the defendants agreed to refrain from doing any of the acts charged in the indictment. This means that the violations were in substance admitted by the persons and concerns under indictments.

In June, 1940, the Federal Attorney General, in co-operation with Mayor LaGuardia of New York City, started an investigation in New York similar to the one which led to the indictments in Chicago. The abuses in New York are similar in character to those which provoked the indictments in Chicago. Many corporations and individuals who admitted guilt in Chicago, and promised to refrain, are factors in the New York system.

Three months later we were told in the press that the investigation had been deferred "because of lack of sufficient evidence to proceed at the moment." It was assumed from the time and nature of the announcement that the investigators were called off. Mr. Thurman Arnold who was in charge of the investigation recently advised us, however, that the investigation will be continued until full and complete information concerning the operation of the New York Metropolitan market has been secured. He said it is probable that the investigation will take several months longer in view of the magnitude of the task.

CHAPTER XXXII

DESPERATE FARMERS REBEL

Dealers Flout Agreement.

During 1939 farmers were suffering from the effects of the severest and most prolonged early summer drouth on record. The ground was parched, there was scant growth in the meadows, crops were short or failed to grow and pastures were burned up. Farmers were obliged to feed fodder intended for the winter. Wells and springs were dry and the flow of milk fell off. With the low prices farmers grew desperate. The leaders of the milk system said they could not increase the price to producers until application was made by the Producers' Bargaining Agency. Then the application would have to be acted on in Washington and Albany. If approved, a new agreement would be drawn, country hearings called and a referendum held to find out if 75 per cent of the producers wanted an increased price, and a vote of the dealers taken to see if 50 per cent of the dealers agreed to it. Notices would have to be given in each step of the proceedings, so if the cows didn't dry up and farmers hadn't starved to death, the price could be raised in about two months and farmers would get the extra money about a month later.

The Dairy Farmers' Union, headed by Archie Wright, consisted of about 15,000 members. They called a strike on August 15, 1939, and demanded $2.35 per cwt. as a flat price—no classification. The leaders of the Big-3 snickered. They had most of the producers tied up in contracts of their system. They were, the leaders said, satisfied patrons. But bound or unbound these producers were desperate. They threw the leaders' previous arguments back in their faces: "Nothing could be worse than existing conditions, and the change might be better." They joined the strike and held their milk.

Some of the Big-3's plants went dry. In some cases the local League officers joined the strike. Farmers picketed the plants to urge their neighbors to hold their milk. Some did but others speeded their trucks through the picket lines. One farmer-picket was run down and died later from the injuries. The Governor sent armed troopers and sheriffs' posses to the plants. Some shots were fired. Dealers set up the false claim that the C.I.O. Labor Union picketed the plants, not the farmers. The C.I.O. gave counsel to the farm leaders and a contribution to the strike fund, but no pickets.

New York City was using 4,400,000 quarts of fluid milk daily. In a week it was short 2,400,000 quarts. The strike was gaining daily. Mayor LaGuardia called a conference of dealers and leaders who favored the strike and those who opposed. The dealer-leaders refused any concessions until Mayor LaGuardia put the heat of his position into the argument. Then a compromise was reached, the dealers agreeing to pay $2.15 for all milk from August 25th to October 31st. The farm leaders agreed to accept that price and terms.

It was a closed conference including Mayor LaGuardia, as arbitrator, three experts representing the three big dealers, both bargaining agencies and the Federal-State officials. Three leaders represented the farmers. Others were turned back at the door.

After the agreement had been reached, the dealers began to use their skill and craft to save their faces. Where and how to get the money to pay for the milk was their concern. They were not asked where or how farmers would get the cost of production; that was the farmers' problem. But the astute dealers proceeded to make up a budget. They proposed to increase the consumers' price thirty-five cents a cwt. and estimated that they could get the extra amount needed out of certain other classes of milk. Apparently it did not furnish enough money to pay in full. Mayor LaGuardia and Mr. Wright insisted that all dealers pay $2.15 per cwt. as agreed. The smaller dealers did. The Big-3 did not.

Whether the dealers worked their budget formula into the written memorandum after the verbal compromise was reached is not known. It was not published. But according to the Big-3, if a farmer agreed to pay $100 for a harvester, estimating that his crop of wheat would be 100 bushels and the price a dollar a bushel, but finding later that the yield was only ninety bushels, he could settle for the harvester for $90. The Big-3 said it was up to the Mayor. Commissioner Noyes issued a bulletin saying the Mayor was responsible for the farmers' disappointment and loss but that, if the equalization claims had been paid in full to the administrator for January and February, 1939, the September bills could be paid in full.

In that view the Commissioner adopted a bit of pure sophistry. The debtors and creditors affected by the January and February defaults were not the debtors and creditors in the September default. Farmers were the losers in both cases, but not the same individual farmers; and, if the first default had peen paid, the second default by an entirely different debtor to an entirely different creditor would be still due.

Mayor LaGuardia, however, demanded a full settlement. The big dealers resisted for a year and then, at the Mayor's insistence, paid up the balance due.

This is a typical example of the way dairy farmers have been "gypped" by milk dealers, leaders and rulers in the sale and distribution of their milk.

The Farmers' 13-cent Dollar.

The Federal-State régime resumed operations July 1, 1939 after the United States Supreme Court decision. The effect of the severe Spring and Summer drouth on the cost of producing milk for the whole year was evident to every one. The restored monopoly announced a price of $1.50 per cwt. to producers. It did not cover cost of production. All appeals for a better price failed and the Farmers' Union strike of that year, already described, followed. The monopoly then adopted a new formula to fix a higher price to cover the farmers' extra costs due to the drouth up to May 1, 1940.

The proponents of the system filled the air and the press for the whole period with propaganda comparing the new formula with the prices which they had forced on farmers during April, May and June, as an evidence of their great benefit to producers and to the dairy industry. It so happens, however, that the first nine months after the Order was reinstated correspond month for month with the last nine months of State Control which proved so unsatisfactory that it had to be abolished. The average so-called uniform price, subject to adjustments, for the nine months, July, 1939, to March, 1940, inclusive, was $2.01 per cwt. There was no corresponding pool price in the Metropolitan area during State Control but it was generally conceded that the prices paid by Sheffield Farms Company was as near an average as could be estimated.

The average basic price paid farmers by the Sheffield Farms Company for the last nine months of State Milk Control, July 1, 1936, to March 31, 1937, was $1.92. Hence after all the promises and boasting, the Federal-State régime paid farmers in consideration of drouth losses nine cents per cwt. more than they received under the discredited State Control when cost of production was much less. During these nine months of Federal-State operations, the dealers increased the consumers' price five times, increasing the cost from twelve cents a quart to fifteen cents. The average price for the period was $14\frac{1}{2}$ cents a quart or $6.81 a cwt.

The average price to consumers during the last nine months of State Control was thirteen cents a quart, or $6.11 a cwt., seventy cents a cwt. less than the average under the Federal-State Orders. Out of the seventy cents extra collected from consumers, dealers paid farmers nine cents per 100 pounds and paid themselves sixty-one cents per 100 pounds in extra profits. Expressed in another way, for every dollar paid by the consumers for drouth relief, farmers got $12\frac{2}{3}$ cents and the milk barons got $87\frac{1}{3}$ cents. Farmers were paid with thirteen-cent dollars.

The table in one of the previous chapters showed the price juggling during the first year of the Rogers-Allen Law under the Bargaining Agency set-up with a net gain to dealers for that year of 66.5 cents per cwt., a net gain to the consumer of 23.5 cents per cwt. and a net loss to the producer of 90 cents per cwt. This

268 SEVEN DECADES OF MILK

price juggling by dealers has not diminished under the Federal-State Order. It has continued unabated. The following table brings the record down to November, 1940:

Price Change	Milk Gain cwt.	Dealers' Loss cwt.	Price to Producers cwt.	Price to Consumers quart
Dealers' net gain 4/1/37-3/31/38 (as previously itemized)	0.665	.....	.....	....
On September 1, price to producers raised 45¢; price to consumers raised 58.7¢	0.137	.....	2.45	13¾
On March 19, 1939, price to consumers reduced $1.087; no Class 1 price to producers	.....	.....	.....	11½
On July 1 price to producers fixed at $2.00, 45¢ less than previous Class 1 price; price to consumers raised 70.5¢, but still 35¢ lower than prior to March 19	0.10	.....	2.00	13
On August 1, price to producers raised 25¢; price to consumers raised 23½¢	.....	0.015	2.25	13½
On August 25, price to producers raised 35¢; price to consumers raised 35¢	.....	.....	2.60	14¼
On October 1, price to producers raised 22¢; price to consumers raised 23½¢	0.015	.....	2.82	14¾
On November 8, price to consumers raised 11.7¢	0.117	.....	2.82 ·	15
On May 1, 1940, price to producers reduced 37¢; price to consumers reduced 23½¢	0.135	.....	2.45	14½
On November 1, 1940, price to producers increased 20¢; price to consumers increased 35¢	0.15	.....	2.65	15¼
	$1.319	0.015		

Thus, under the Federal Order-Bargaining Agency régime, the dealer spread has increased in the past 3½ years $1.304 per cwt., or 2.7 cents per quart. There was no official price announced from March 31, 1937, when State Control ended, to July 1, 1937, when the Bargaining Agencies started to fix prices. Again, there were no official prices from March 28, 1938, when the Bargaining Agency set-up collapsed, until September 1, 1938, when the Federal-State Orders went into effect. Nor were there any official prices from March 19, 1939, when the Orders were disqualified, until July 1, 1939, when they were reinstated. The dealers, however, fixed prices for both producers and consumers during these periods just as they did before and just as they have done since.

Antics of Desperation.

When the Federal and State Orders were invalidated by Judge Cooper and Judge Bergan in February, 1939, the dealers proposed to continue the policies, formula and methods of marketing under agreements among themselves in defiance of the courts. They fixed up a contract to become effective if 90 per cent of the distributors signed it by March 14th. They reported that only 88.1 per cent signed and declared it a failure; being 1.9 per cent short.

The directors of the Dairymen's League then held a meeting in Syracuse on March 16th and adopted a resolution saying that it would sell milk on a competitive (cut-price) basis. This was reported in the *Herald-Tribune* on the morning of March 17th. The retail price then was 13¾ cents a quart for Grade B milk and 17¾ cents for Grade A. On the next day William H. Marcussen, president of Borden Farm Products, announced a cut in its retail price of 2¼ cents a quart to consumers, explaining that the direct cause was the resolution adopted by the League and the fact that the League was negotiating new prices with dealers.

On March 18th, the *Herald-Tribune* published the following statement:

"Sheffield Farms Company followed the lead of Borden's Farm Products, Inc., yesterday when it cut the price. . . . Borden's announcement was made Friday night.

"The leading independent dealers called a hasty meeting and announced that they would meet the price reductions of the two major firms. They did so grudgingly, however, contending that the drastic price cut constituted a maneuver to freeze small dealers out of the field.

"After a long and stormy meeting, the independent dealers' organization issued a statement which said, in part:

" 'The independent dealers are opposed to starvation pay for farmers. We are prepared to pay a fair price to farmers, but the Milk Trust has always prevented this by starting price wars. The present case is not new. Of course we must meet the new prices.' "

And again on March 23rd:

"Borden's Farm Products, Inc., which gets its milk from the Dairymen's League, instituted the two and a quarter cent slash and was followed by Sheffield and the independents."

The following week Mr. Marcussen stated in a signed advertisement published all over the State that "certain elements" had caused all the trouble and that Borden's was losing one cent on every dollar's worth of milk sales. (He made a somewhat similar statement in December, 1937, but Attorney General Bennett showed that at that time his profit was 1½ cents a quart.) After the 2¼ cent slash the retail price was 11½ cents a quart. The Borden-League return to the farmer for April, the first month after the slash, was two cents a quart, or ninety-four cents a cwt. Reports from up-state were as low as seventy-four cents per cwt. For the month of May the Dairymen's League paid net eighty-seven cents a cwt. For that same cwt., or forty-seven quarts, the consumer paid Borden $5.40.

For the third time in the period from April 1, 1937, to July 1, 1939, the Big-3 had deliberately provoked a price war in the city retail trade, as an excuse for paying the farmer a beggarly blended price for all the milk so that they could again tell the farmer that he could be no worse off than now and he might as well try them again for the chance that their new scheme might be better. If the leaders of the Dairymen's League had any concern for dairy farmers at that time and if they were free to act, it was their opportunity to insist on continuing the prices then in effect. If they had done so, their own producers as well as every independent producer and dealer would have done the same. They said so.

The truth is they were not free. They were one of the Big-3 dominating the milk trust. They were the spokesmen of the other two members, who stood there convicted by Judge Cooper of deceit, corruption and fraud, on evidence which has never been challenged or revoked. Instead of demanding that the trust continue existing prices to producers and consumers, the League and its Bargaining Agency first tried to frighten farmers into believing that they faced a great calamity and the only way farmers could survive was to sign all their rights and interests over to the racketeers. Failing in this they deliberately struck dairy farmers as cruel and as disheartening a blow as any bombs Hitler ever showered on the peaceful homes of his helpless victims.

From Thrift to Gross Extravagance.

The farmer first sold milk direct to the local consumer and kept 100 per cent of the dollar. He then shipped to the city and employed an agent. The agent became a dealer and the dealers banded together to bargain with each producer separately. This led to the farm co-operative. In time the co-operatives fell under the control of the dealers who pressed their advantage almost to a farm collapse, creating an emergency. This brought politics into the picture with a mixture of State and dealer control. The collapse of this adventure was followed by the bargaining agency sham set-up under authority of the Rogers-Allen Law. The quick failure of that obviously bogus creation was made the excuse for the Federal-State régime, which is trying to sell monopoly control to dairy farmers, and trying in vain to make the milk producer think he likes it, at $20,000,000 per annum at least.

Some idea of the complications and extravagances of the system since the farmers' first simple plan may be seen in the list of agencies employed in wholesaling the milk, each taking a toll out of it.

1. The Dairymen's League Co-operative Association
2. Dairy-League Co-operative Corporation
3. League Regional Committees
4. League County Committees
5. Local League Branches
6. Metropolitan Co-operative Milk Producers' Bargaining Agency
7. Metropolitan Co-operative Milk Distributors' Bargaining Agency
8. Eastern Milk Producers' Association
9. Dairy Farmers' Information Council (Dairy Farmers' Digest; Milk Tattler)
10. Division of Milk Publicity (New York)

11. N. Y. Department of Agriculture
12. U. S. Department of Agriculture

These agencies exist for wholesaling milk to dealers who retail milk to families or sell in cases or cans to stores, hotels, and other institutions. The exact cost of these agencies is not known but it is paid by milk producers,

These agencies have six periodicals to spread propaganda for the system:

(1) The Dairyman's League News
(2) Eastern Milk Producer
(3) Metropolitan Milk Producers' News
(4) Dairy Farmers' Digest
(5) The Milk Shed Tattler
(6) American Agriculturist

The first on the list is printed weekly. The producers for the League and the Borden Company have $1 a year taken out of their milk returns as a subscription whether they want it or not. It is admitted to pound postal rates, but it has in addition a deficit of as much as $40,000 a year charged to producers as expense.

The second on the list is a Sheffield producers' organ with like mail privileges. I am not fully informed about its financing but in one way or another, the cost comes out of the producers.

The third, fourth and fifth on the list have no standing as publications. Our information is that the circulation is gratuitous. They are purely propaganda sheets. Their printing and mailing and postage reduce farmers' milk returns. They are distributed monthly.

The last on the list gets its share out of the milk system as a subsidy in gross lumps. By courtesy, that is called advertising patronage. It runs into thousands of dollars a year.

For the most part producers for the system pay the cost of all this publicity. The worst thing about it is that the farmer has no choice. He must pay for it whether or not he wants to do so. The cost is taken out of his milk returns without his voluntary consent, and in many cases against his will. They complain bitterly, but they just cannot avoid the imposition.

Failure Admitted From Within.

In the 70 years from 1870 to 1940, seven corporations controlled by milk dealers have dominated the distribution of milk in the New York milkshed, and fixed prices for both producers and consumers, except for about six months in 1882 and a full year following the October 1916 farmers' united revolt. The names of the corporations are:—

1. The Milk Exchange, Ltd.
2. The Consolidated Milk Exchange, Ltd.
3. The Dairymen's League Co-operative Ass'n., Inc.
4. The Sheffield Producers Co-operative Ass'n., Inc.
5. State Milk Control.
6. The Metropolitan Producers Bargaining Agency, Inc.
7. The Metropolitan Distributors Bargaining Agency, Inc.
8. The Federal-State Orders.

The fifth and eighth may not strictly be called corporations, but both were created by law.

The first, organized in 1882, was the prototype of all the others. The alleged purpose was to create a legal organization in which farmers and dealers would together determine a fair price to pay producers for milk. In the meeting to organize, the dealers demanded a majority of the directorate, and farmers "took a walk." The Exchange was organized with a strong majority of dealers, and a minority of subsidized stooges, and small dealers. This dominating majority and subservient minority have characterized all the organizations that followed. In 1895 the Supreme Court found The Milk Exchange, Ltd. unlawful and fraudulent, and annulled its charter. Its successor, a foreign corporation, was also attacked by the Attorney General in the courts as unlawful and fraudulent and cancelled its permit to do business in New York State.

The six successors were organized on the same formula and for the same purpose. The methods have become more refined, subtle and effective. For 50 years dairymen had the benefit of basic laws. Their rights were recognized by the courts. Then, assuming that bills proposed by the alleged farm leaders would be fair to milk producers, the Legislature enacted laws which actually deprived farmers of their inherent rights and clothed milk dealers and their stooges with tyrannical power over milk producers. Later on, a legislative bloc of members in both branches was formed in the Legislature. The dealers had a registered lobbyist at Albany and the record of milk legislation indicates the power of the milk barons. The milk racket was triumphant. Producers were made subservient to the milk barons, and finally the milk monopoly was legally sanctioned by the Federal-State Orders. After the Dairymen's League Co-operative Association Inc. had been in alliance with the Borden Company for nearly 20 years, Fred H. Sexauer, president of the Dairymen's League, was quoted as saying that

"farm buildings are falling down, land fertility has been destroyed, homes and furniture are worn out, the health of farm people is neglected. They can't afford medical care. Farmers have no agency to assure them of a fair price for milk."

Again I quote Mr. Sexauer from the daily press of the same year:

"The industry is 'honeycombed' from top to bottom, honeycombed with inefficiency on roads, in country handling, in city delivery, honeycombed with political and personal ambitions, with personal pique, with newspapers using perverted milk news as a subscription builder, with dealers crowding roads with men to destroy dairy organizations. The strategy of milk buyers 'has been to keep farmers competing for an outlet for their milk' so they can buy at low cost and undersell competition. Too many milk dealers want to make their money buying milk."

This was all he could say for the alliance and the system that he had painted for two decades as the most perfect in the world, and as a boon to dairy farmers. I do not know what inspired these frank and truthful statements. The time was after the publication of Attorney General Bennett's Report, and the Ernst and Ernst Audit, with their damaging revelations of rebates to dealers, burned vouchers, intrigue with Borden on audits, and annual expense of $20,500,000 taken out of League producers' milk returns.

It was reported at the time that the men who had the last word in the Borden Company were inclined to withdraw from the League alliance. It was said the hesitation was due to difficulties in finding a way to make up the rebate of $3,000,000 a year derived from the alliance. That would seem to imply that Borden received its full share of the six million dollars in rebates reported as paid back to dealers the year of the audit, but the testimony is that Borden had received the rebate from 1922 on. But definite information is not available. The official auditors were unable to find who got the $6,000,000 rebate the year of the audit.

State officials at Albany, including Governor Lehman and Agriculture Commissioner Noyes, have admitted that dairy farmers should be free to negotiate the price and sale of the milk they produced; but they insist that farmers fail to agree and consequently the state assumes the obligation. My answer is that farmers when left to themselves did agree practically to a man in 1916, and were successful, and that a greater percentage of them demand freedom to conduct their own business than the percentage of them that are forced by economic pressure to approve the state authority. Sixty per cent of them joined the strike in 1939 for an increased price. No such number has ever approved the monopoly.

Mr. Noyes said he would not be satisfied with a blended price of less than $2.00 a cwt. for milk. He expressed the judgment that the charter, designed as a means by which farmers control their own business, would not work. The average prices paid farmers under the system which he approves for $3\frac{1}{2}$ years are $1.47 a cwt. paid by Dairymen's League, and $1.685 paid by Sheffield Farms, since April 1, 1937 to November 1, 1940. I submit that the political system does not work for either producers or consumers but the dealers' gains as shown above seem to be satisfactory to the savants of the monopoly.

CHAPTER XXXIII

THREE OFFICIAL REPORTS

The Federal Trade Commission Report.

Congressmen Fred W. Sisson of Utica, New York, and Herman P. Koppleman, of Connecticut, worked diligently with friends of dairymen of New York and other states for an investigation of the dairy interests by the Federal Trade Commission. Authorized by a resolution of Congress on June 15, 1934, investigations were conducted in Philadelphia, Baltimore, Connecticut, and Boston. The appropriation was exhausted without touching New York. New York members appealed for a new appropriation and got it. With this the investigators went to Chicago, St. Louis, Cincinnati, St. Paul, and Minneapolis. By the time these fields were covered, the second appropriation was nearly exhausted. Acknowledging that "it is one of the most important milk markets," the Commission stated that, "due to lack of available funds, it was not possible for the Commission to make a detailed investigation of conditions in the New York Milk Shed."

It was generally understood in dairy circles that the New York milk ring had backing from the highest powers in Washington, that the Commission regarded the New York situation as a "hot spot" and that it had no disposition to "burn its fingers." The Dairymen's League, it was said, had engaged a special lawyer who enjoyed friendly relations with the Commission's offices. The Borden Company was represented, it was said, by an ex-Ambassador. National Dairy interests were looked after by a high class legal firm of New York City.

The Federal Trade Commission published a Report on January 5, 1937 which contained incomplete information "gathered during the investigation of the large dairy companies and farmers' cooperative organizations with headquarters in New York City." This report is known as House Document No. 95.

The Report was first released in mimeographed form. The Dairymen's League printed it in pamphlet form much resembling the documents containing the Commission's reports, with a personal imprint on the first cover page and with emphasis on the text paragraphs which the League officials interpreted as favorable to the management of the Dairymen's League. I received one of these which had been delivered by mail under a "frank" of the Federal Trade Commission. This was clearly an illegal use of the "frank." My editorial on the subject was republished. It created some correspondence between citizens and myself and the Commission. It is my information that the reprint of the report displeased the Commission and that the edition was not fully distributed.

A Little Bit Too Innocent.

In some respects the Report did seem to have a League bias. One statement which the League officials stressed as an unqualified indorsement said that an examination of the records subsequent to 1934 was made and

> "no evidence was found indicating that there was any special arrangement between the two organizations whereby the Borden Company received any concessions not shown in this report."

This seemed clearly to mean that concessions were shown in the Report. In fact, the three preceding pages plainly revealed many such special concessions in the sale of milk. The Commission had been directed to public testimony by a former President of the League to the effect that Borden's had a secret rebate from the first. Moreover, the Commission's Report showed two Dairymen's League annual statements containing disguised deductions for 1934 and 1935, which the Ernst & Ernst audit later showed in a similar statement for 1937 to be rebates to dealers of more than $6,000,000 for that particular year. It would be a very innocent investigator who would assume that Borden's did not get their share of such rebates, if not the whole.

Another statement in the Report which gave the League officials much comfort was the failure of the investigators to find proof of any relations between Dairymen's League and the Borden Company other than that of "vendor and vendee."

The investigation was ordered two years before any files were searched for records to show Borden's control of the League. It had been discussed in Congress two years before it was passed. No records previous to 1934 were examined. Did the Commission expect that the officials would deliberately keep evidence in their files to convict themselves? That seems just a little bit too innocent.

The League's Marketing Cost.

In the Federal Trade Report were published the incomplete annual statements of the Dairymen's League Co-operative Association ending March 31 of each year from 1930 to 1936. In February, 1938, Attorney General Bennett published a corresponding statement for the year 1937. In 1938 Commissioner of Agriculture Noyes employed Ernst & Ernst, certified public accountants, to make an audit of milk dealer and co-operative books. The figures in the following table, except the computations in the columns headed "Cost of Marketing" and "Expense per cwt.," are taken from these reports. For the first time in 16 years this table gives producers an account of their cost for selling their milk. The reports of the Federal Trade Commission, Attorney General Bennett, and Commissioner of Agriculture Noyes have made this possible for the period from 1930 to 1937. The table shows the annual selling cost for this period in bulk and the selling cost per hundred pounds of milk on an average from year to year for 16 years. The table could not be made up from any one or any two of the reports. All three were required to complete the data. Of course, the auditors knew nothing about the correctness of weights or fat tests or bacteria counts.

These figures take no account of grades or classes of milk. They do not give details to reveal to farmers what their money was spent for, nor show who got it. The table simply takes the amount of the "Gross Sales" as reported and subtracts the amount "Paid Producers." The difference is the "Cost of Marketing," and dividing this item by the number of hundredweights reveals the selling "Expense per cwt." The data in the table from 1922 to 1929 is not official. It has been taken from statements and text published in the *Dairymen's League News*, the official organ of the management.

The table follows:

COST OF WHOLESALING MILK THROUGH THE OFFICIAL GROUP
OF THE DAIRYMEN'S LEAGUE

Year	Pounds of Milk	Gross Sales	Paid Producers During the Year	Cost of Marketing	Expense Per Cwt.
1937	2,557,983,280	$69,653,677.99	$48,117,609.01	$21,536,058.98	.84
1936	2,376,312,175	57,275,774.98	41,624,433.16	15,651,341.82	.66
1935	2,487,879,074	58,218,112.07	40,384,528.25	17,833,583.82	.71
1934	2,673,758,199	55,454,822.85	37,337,480.04	18,117,342.81	.49
1933	3,123,189,866	55,140,146.78	34,535,830.27	20,604,316.51	.66
1932	3,124,116,430	70,156,911.17	49,198,721.27	20,958,189.90	.67
1931	2,793,866,326	80,165,183.60	59,321,511.74	20,843,671.86	.74
1930	2,623,684,485	84,473,526.62	65,786,563.17	18,686,963.45	.71
1929	2,484,941,739	85,741,658.27	62,202,694.03	23,538,964.24	.94
1928	2,420,384,585	82,501,310.00	60,491,773.49	20,172,253.91	.83
1927	2,224,220,066	73,845,097.22	52,597,225.36	21,247,871.86	.95
1926	2,376,312,176	66,699,331.46	52,338,887.38	14,360,444.08	.60
1925	2,478,879,074	65,067,864.29	46,737,875.19	18,329,989.10	.73
1924	3,095,706,972	75,131,869.50	59,607,993.80	15,523,875.70	.50
1923	3,735,998,307	82,130,902.17	63,421,411.47	18,709,490.70	.50
1922	2,976,381,692	61,862,075.05	47,775,440.95	14,086,634.10	.47

The data since 1938 is not available to the public or to me.

The "Gross Sales" does not represent the consumers' cost. It is the wholesale cost to dealers. The dealers sold milk to consumers or to stores and the consumers paid the cost of retail distribution and dealers' profits, in addition to the wholesale price.

This is a staggering cost for the sale of milk. For the last decade, 1928 to 1938, the annual cost has averaged just about $20,000,000. This is a yearly average, per member, of $666 and, per one hundred pounds of milk at wholesale, of 76.5 cents, compared with the cost of 1 cent a hundred pounds from 1916 to 1922, when milk prices were relatively higher. The farmer usually gave up one can of milk of every three to pay the cost of selling the other two cans.

Before pooling and classification were adopted, the farmer paid one cent a cwt. as a selling commission on which the League accumulated a surplus. Any broker would take this volume of milk today to sell on the one cent a cwt. commission and make a fortune. To be exact, his commission would amount to better than a quarter of a million dollars every year.

That is not all the discouraging part of the record. The price of milk and also the purchasing power of the milk dollar steadily declined from the time the Dairymen's League Co-operative Association began to operate down through the 1930's. In 1921, the last year that milk was sold at a flat price and the farmer knew the price before it left the farm, the average price per hundred pounds for the year was $2.46 as compared with $1.67 for 1937. This is a drop of 59 cents a cwt. The average price as reported from January 1, to June 30, 1938 was $1.33, and the farmer did not always get all the price reported.

No Farm Parity.

The government keeps a record of the average annual prices of things farmers buy. If a dairy farmer paid $100 for a given amount of such supplies in 1912, 1921, 1924, 1932 and 1937 and sold enough milk each of these years to pay for the farm supplies, the pounds of milk required to pay for the same supplies each given year would be as follows:

```
1912: Milk required ............ 6289 lbs.
1921:  "      "     ............ 6720  "
1924:  "      "     ............ 7451  "
1932:  "      "     ............10000  "
1937:  "      "     ............ 7800  "
```

My testimony before the Pitcher Committee clearly demonstrated that while the price to the farmer decreased, the profits to dealers sharply increased. Their spread increased both in volume of money and in the purchasing price of their dollar. In 1912 the government quotation was $1.59 per cwt. to New York producers, or 3.38 cents a quart. The cost of a quart bottle of milk delivered to the consumer's door in New York City was nine cents. Similar computations for the other years give the following results:

```
1912: Dealers' spread per quart ........ 5.62  cents
1915:    "        "    "   "    ........ 5.2     "
1924:    "        "    "   "    ........ 9.79    "
1931:    "        "    "   "    ........10.4     "
1932:    "        "    "   "    ........11.84    "
1934:    "        "    "   "    ........ 7.322   "
1937:    "        "    "   "    ........ 8.29    "
```

During the later years, the additional volume of pint bottles, Grade A, Vitamin D, and Certified milk sold at increased prices would show a still further increase in the actual spread.

Profits in Grade A and Vitamin D.

Until Grade A milk was abolished in September 1940, dealers charged consumers a premium of three cents a quart for Grade A milk over the Grade B price. They paid farmers a sliding scale from four cents to eighty cents a hundred pounds of milk, depending on the bacterial count. I have seen returns to the producer showing that the consumer received 6/100 of one cent per quart as a Grade A premium. The producer received one cent premium on 16 2/3 quarts. The dealer received a premium of fifty cents on the same 16 2/3 quarts. This was a profit of forty-nine cents on one cent, or a profit of 4900 per cent on the Grade A premium.

The profits in Vitamin D, from cod liver oil, are fabulous. The residue of the oil is used for making hair tonics and soap. The extract is sold in small cans at $2.50 each. The contents of a can is sufficient for 5,000 quarts of milk. The extract is simply poured into the milk. The cost is one-half a mill (.0005 cent) a quart. The milk is then sold for one cent extra per quart. Twenty quarts are treated at a cost of one cent. The profit is 19 cents on 20 quarts. This is a 1900 per cent profit. The New York law holds any milk containing any foreign substance to be adulterated. In this case the size of the profit seems to have neutralized the adulteration.

Complaints From Producers.

The Federal Trade Commission found many letters from farmers in the files complaining of lack of information about the League activities and Borden's connection. Some examples were printed in the Report.

W. Krause wanted to know why he got three cents a quart for milk when Borden's got 10 cents a quart for it, and "how much Borden's paid the Dairymen's League for 4.1 per cent milk."

The answer did not tell him.

V. W. Utts, Cassadaga, New York, complained of low price. He thought butter and cheese would pay more. Mr. Sexauer tried to persuade him that the price was all that could be expected under the circumstances.

J. H. L. Todd, Warwick, New York, protested the closing of a plant and asked if Borden's controlled the League. Mr. Sexauer argued that he had just closed a plant where farmers would now deliver at a distance of 9 miles from the closed plant, while the Warwick farmers had only to deliver 3½ miles from the closed plant. He said Borden's did not control the League.

George W. Medbury, Rockdale, New York, wrote that one of the big troubles charged against the League is this: "They are owned and controlled by Borden's." H. H. Rathbun, an official, replied with another question: "Have you or anyone else faith in the directors you elect?"

Mrs. Edward Kelly, Friendship, New York, asked: "Why cannot the members of the League find out the price paid for their milk when it is taken to the Borden plant? The superintendent doesn't know and nobody can find out even from our directors." Mr. Sexauer wrote in reply in part that Borden "pays for the milk on the basis on which it is used." He said, "Their books are audited to determine in just what form Borden sold this milk, whether in the form of fluid milk, cream or condensed milk or butter and cheese." (After Attorney General Bennett found that no such audit had ever been made, Mr. Sexauer then admitted that he made no audit.)

On March 11, 1935, M. L. Zimmer, Covington, Pa. wrote in

part: "I would like to know what progress we have made since 1932. Every bit of legislation that has tried to be put through to help the farmer, the Dairymen's League has been right there for the big dealer—Borden." Mr. Sexauer said in his defensive reply: "I cannot believe that down in your heart after 14 years you believe that your association has not been working in the interest of producers." (The record of legislation in New York State down to 1940 confirms Mr. Zimmer's statement.)

Country Plant Operations.

The Commission accepted the explanation of Mr. Sexauer on the benefits of closing country plants, and on volume differentials. The Report copies a letter from the president of the League dated October 5, 1934 to Mrs. John Winchester, Meshoppen, Pa. The letter says: "By the elimination of some plants, the association has saved approximately 18 cents per hundred on all milk that is handled. This saving has been directly returned to members through the pool account."

There is also a saving of about twenty-three cents a cwt. on milk delivered in tanks, cars and trucks. This, with the 18 cents saving in volume plants, should add nearly a cent a quart to farm returns. Does the farmer get it? Ask him. He knows. My answer is, he does not.

Again, in making returns to producers, up to a short time ago, Mr. Sexauer refers to a fifteen cent average differential "which varies both according to locality and differentials earned by individual producers." This fifteen cents was deducted from the pool returns, leaving what was called "the gross pool return." I have never been able fully to understand these returns and have never found anyone who did. If all returns for milk are put in a pool, it seems elementary to say that when some producers get six, eight, ten or twelve cent differentials according to the accidental size of their plants, and others get nothing because their plants are accidentally smaller, then those who get the least or none are helping pay those who get more. When the differentials are fifteen to eighteen cents, some get more and others get less.

The destruction of so many local plants requires that some

milk must be trucked as much as twenty miles or more. This involves a trucking charge of ten, fifteen or twenty cents per cwt., according to distance, and is charged back to the farmer after his price per cwt. is computed. In other words, the farmer's net return is short the cost of the cartage to the plant; but the volume differential of six, eight, ten or twelve cents is in the price to improve his morale. I know farmers who formerly delivered direct to their local plants. Now they pay twenty cents per cwt. for cartage to a volume plant. They lose any way they look at it, and they don't like it.

But Mr. Sexauer was not candid with the Commission. The purpose of the volume plant was not for economy. The closing of plants did not as a whole benefit farmers. On the contrary it caused great losses. Most dealers were paying more. Farmers were quitting the League. Borden's was growing restless. At one time Borden's broke out and bought a volume of milk direct from farmers and paid more than the League paid its farmers. The League officials bought the plants with farmers' money giving IOU's for it. Their policy was to control the only plant within a shipping area so that the producers in that area would be obliged to sell to their plant and sign the contract and like it. Their control of the plants would hold producers and the power to fix low prices and rebates would hold Borden's and other dealers. That is the plan that cost League producers $21,500,000 in 1937 for selling their milk to wholesale buyers.

Dealers Cheat Producers.

In another Federal Trade Commission Report, (House Document No. 152) it is disclosed how certain dealers in Philadelphia and Connecticut cheated milk producers out of more than $600,000 in the year 1934. This loss was estimated for the year from a study of the dealers' books for one month. The losses to farmers resulted from the dealers' practice of underpayments (1) on milk sold under the utilization basis, (2) by dealers buying on flat prices, (3) on milk sold as class one, and (4) on profits in hauling milk to city stations.

By buying out the business of independent distributors the big dealer companies in both areas have tended to monopolize the markets. In six years National Dairy received from its two subsidiaries in Philadelphia $27,500,000, or more than 70 per cent of the cost of acquiring them.

Mr. W. A. Ayres of the Federal Trade Commission also explained that the Philadelphia dealers reduced the blended price to farmers by importing cream from outside the milkshed to create a surplus in Philadelphia. The base plan was modified from time to time without correcting the abuses. He further states:

"Even under the Agricultural Adjustment Administration's marketing agreement for the Philadelphia milkshed and subsequently under the Pennsylvania Board of Milk Control obligations, not all of the Philadelphia distributors buying on the base-surplus plan paid producers fluid milk prices for all milk such distributors sold as fluid milk."

Mr. Ayres pointed out that producers were at a disadvantage in the Inter-State Milk Producers' Association, a co-operative association supplying Philadelphia with milk, because the Producers' Association had not adequately protected the interest of producers, and consequently the farmer was at a serious disadvantage in his relations with dealers.

What the Federal Trade Commission Report Did Not Show.

The Federal Trade Report (New York) gave an analysis of the Borden Company and of the National Dairy Products Corporation, including the names and numbers of their subsidiaries. It reviewed the set-up and procedure of the Sheffield Producers' Co-operative Association, Inc. and concluded, after analyzing the contract between the Association and its buyer, that it was dominated by that exclusive buyer, the Sheffield Farms Company.

The Report did not enumerate the subsidiaries of the Dairymen's League nor show whether or not they were operated at a profit to producers. It did not analyze the pooling contract between the League and its producers. It did not verify its information that the pooling contract requires farmers to waive their right to an accounting, nor did it call attention to the fact that for fifteen years the League paid its producers from about

30 cents up to one dollar a cwt. less than Sheffield producers received during this period. The ambiguous "no evidence" finding on the Borden-League alliance was most unsatisfactory.

It should be noted that the Federal Trade Commission stated no general investigation could be made at the time because funds were not available. I feel that an error was made in attempting a temporary report of the New York dairy situation. It was then and is now crying for a thorough and impartial investigation by an authority able and willing to go to the roots of the evils in New York State. The Commission has rendered such service to agriculture in other fields. I hope it may yet return and make a complete job in New York.

The Report called attention to the complaints of League producers of the lack of proper information and noted that the "director does not give much of a detailed account of business transacted."

While much needed information is lacking, the publication by the Federal Trade Commission of the seven annual reports of the Dairymen's League, not previously available, shed some light on the problem. Unfortunately, the tables of figures were so muddled that no one but an expert was likely to work out of them the information that most concerns farmers. The Commission did not discover that the item headed, "freight returns and allowances," in these financial reports, really means "rebates or kickbacks" to dealers, as was shown later by the Ernst & Ernst audit, and abundantly verified elsewhere. This oversight led to an error in computing the selling cost to League producers. The Commission gave this cost as 52 cents per cwt. for 1935, and 50.3 cents for 1936, whereas the actual cost was 71 cents for 1935 and 66 cents for 1936. The Ernst & Ernst audit showed the 1937 cost to be 84 cents per cwt.

F.T.C. Recommendations.

In House Document No. 94 accompanying the incomplete report on the New York market, Mr. Ayres expressed some conclusions which may be condensed as follows:

Prices paid producers in most markets were based on quoted prices of butter and cheese in Exchanges where the buyers fixed the prices.

A drop in prices to consumers is followed by a decrease to producers. An increase to producers has been immediately followed by an increase to consumers and in most cases the increase to consumers is more than the increase to producers.

During the depression, while prices paid producers reached a very low point, and prices to consumers were lowered, at the same time the distributor in most instances made a fair margin of profit.

In markets where milk is sold at prices varying with use, the reports of dealers have not been properly verified by thorough audit of distributors' books to determine a correct blended price for the protection of producers.

There is evidence that large distributing companies have granted secret rebates to certain customers including chain stores.

Large dairy-products concerns manufacturing various products and distributing fluid milk in a number of markets, use their purchasing power and the classification plan to depress prices paid to producers for milk. "Such practices are detrimental to the interests of the consuming public."

He further advises that

1. Co-operative organizations be broadened and encouraged. However, milk producers' co-operative associations should be *controlled exclusively by the dairy farmer members*.
2. Each member should have an equal voice in the management and the evil of voting by proxy should not be permitted.
3. Classification and blended prices should not be used unless farmers are in position to have an audit made by their own accountant.
4. Co-operatives should make full and detailed accounting and complete reports.
5. Legislation, designed to provide such requirements, should be enacted.

These recommendations were not underscored in the Dairymen's League reprint of the Report. I have persistently urged all of these recommendations except the qualification in Number 3. No audit could fully eliminate the perils in classification. Classification cannot be justified. Wherever it is used, farmers are cheated. It was adopted for that purpose. If it had failed in that purpose, it would have been discarded by those who invented it.

The State Audit.

Full and detailed accounting and periodic reports have always been implied and expressed in co-operative organization proposals. It is a matter of law that a trustee is required to make an accounting of his trust to his principal or ward. But the Dairymen's League pooling contract contains a provision exempting it from any accounting to its producers. Year after year, Senator Rhoda Fox Graves, of St. Lawrence County, has introduced legislation to require all co-operatives and agents to make proper accountings to their memberships. It has always been opposed and defeated by spokesmen of the League.

The 1937 fight of the New York State Milk Committee, an organization of 18,000 dairy farmers, for a special dairy charter to unify dairy interests, emphasized the importance of an accounting among desired objectives and during that session the Legislature appropriated $50,000 for an audit of the books of milk dealers and dairy co-operative associations by the Commissioner of Agriculture and Markets. Mr. Fred H. Sexauer, President of the Dairymen's League, proposed and argued that books and records of his organization be exempt from the audit. He was overruled.

The money available not being sufficient to audit the books of all dealers and co-operatives, 14 dealers and the Dairymen's League, as the largest co-operative, were selected for an audit covering a period of twenty-one months' operations. Ernst & Ernst, a firm of certified public accountants, were employed by the Commissioner to make the audit. It is my information that all the dealers opened up their books and records and turned the accountants free in their offices to examine any records they desired to see. At first the League management denied the accountants an opportunity to make a satisfactory audit. After an interview with Commissioner Noyes in which a stern alternative was visualized, President Sexauer permitted the auditors to proceed, but not with the fullness and freedom permitted by the other houses audited.

Borden and Sheffield Profits.

The auditors were limited to three and a half months for their examination and in their Report continually pointed to the fact that this period of time was insufficient for them to make as complete a study as the facts demanded. Much of the data obtained was merely from information given to them by companies' officers and employees and no accurate verification of such information could be made within the time allowed. This was particularly true in the reporting on the classified price plan.

Each milk dealer has his own particular accounting method and Ernst & Ernst points out the difficulties inherent in such confusion. They recommended a uniform system of accounting for all dealers.

An analysis has been made of the audit of the books of each dealer. The following is a summary of such analysis:

Under the circumstances this audit of the dealers' books revealed little of importance except that the records of the big companies showed heavy capitalization and extraordinary surpluses. Charges against earnings included heavy depreciations and write-offs, duplicating salaries in the case of Borden and Sheffield subsidiary companies, unitemized and excessive expenses and supplies, and dividend payments disguised as notes. This type of bookkeeping tended to reduce the net earnings or profits and dividends. For example, Borden's Farm Products, the Borden subsidiary which buys all its milk from Dairymen's League, showed net earnings of $1,739,704.54 in 1936, a return on gross sales of 8.87 per cent. But after the accountants adjusted the accounts to realities, the earnings showed 18.8 per cent instead of the 8.87 per cent as shown by the books.

In like manner, Sheffield Farms Company showed a net profit of $2,229,419.89 for 1936, representing a return on gross sales of 5.75 per cent. The adjustments raised this to 8.27 per cent.

If dairy farmers could pay salaries to themselves, wages, interest, taxes, and all expenses out of their income, and charge off losses and depreciation, they would be satisfied with smaller dividends than these and there would be no farm problem.

Analysis of League Expenses.

The audit covered the financial record of Dairymen's League and its subsidiary companies, wholly owned by the League and operated as branches, as follows: Beakes' Dairy Co., Cloverleaf Creamery, Inc., DeLancey Milk & Cream Co., Inwood Farms, Inc., Miller's White Farms, Inc., Model Dairy, Inc. (formerly Hauk & Schmidt), Ferndale-Nelson Creamery, Inc., and Sacks' Dairy, Inc.

In the following organizations, the League was shown to control more than 51 per cent of the stock ownership: Callicoon Farmers, Inc., Canajoharie Farmers, Inc., El-Cor Dairies, Inc., Ft. Plain Producers, Inc., Pulaski Dairymen, Inc., and St. Johnsville Farmers, Inc. Many other companies have been purchased and absorbed by the parent company and its subsidiaries.

The figures for the fiscal year ending March 31, 1937 are as follows:

INCOME

Gross Sales (on 25,579,832.8 lbs. of milk) $69,653,667.99

EXPENSES

Allowances and Returns (Price Adjustments to Dealers)	$ 6,055,167.43
Wages and Salaries	3,890,742.60
Salaries — Officers	43,300.08
Fuel, Power, Water, Ice, Refrigeration	909,876.80
Containers and Sundry Supplies..	725,460.42
Hauling — Interplant	241,996.88
Sundry Transportation and Warehouse Charges	107,994.60
Freight, Express and Cartage from Plants to Branches	906,662.63
Freight, Express and Cartage from Plants to Warehouses to Customers	1,033,039.59
Hauling from Branches to Customers	129,033.66
Auto Expense, including Depreciation, Wages, Repairs, Insurance, Gas, Oil	693,703.61
Taxes and Licenses	195,888.09
Depreciation	793,894.53
Repairs and Maintenance	298,327.98

Rent	142,344.14
Insurance	171,726.56
Advertising	107,231.88
Tel. and Tel., Staty., Prtg. and Postage	163,940.51
Office Supplies and Expenses	48,264.96
Traveling Expenses	171,086.23
Professional Services — Legal, Auditing, Etc.	109,805.15
Meetings	56,017.60
General Membership	59,271.65
Provision for Bad Debts	252,413.05
Other Expenses	419,253.80
Veterinary and Dairy Service	*12,101.26
Expense Transfers	*18,935.06
Handling and Processing Charges	662,183.88
Revaluation and Write-off	491,080.12
Loss on Leased Plants	6,610.99
Loss on Dairymen's League News	7,534.61
Other Deductions (net)	250,813.88
Cost of Cream, Milk Products and other Products Purchased	742,153.47
Distribution to Locals	76,881.41
Certificates of Indebtedness	1,793,918.91
Totals	$21,726,585.38
Less Increase in Inventory	190,526.40
TOTAL EXPENSES	$21,536,058.98
Paid Producers	48,117,609.01
	$69,653,667.99

* Credit.

Both income and outgo figures are lamentably lacking in details. Almost from the start the League published monthly reports of expenses in a form of which the following is typical:

Total administrative expense amounts per hundred pounds to . .03

Deduction is being made to be distributed to local associations, to cover local expenses per hundred pounds .... .002

Deduction is being made to be distributed to sub-district organizations to cover expenses per hundred pounds .... .001

Deduction is being made to create insurance fund to cover losses by uncollectible accounts and other losses .... .005

Deduction is being made to cover depreciation in value of fixed assets due to decline in cost of building and equipping plants and to cover good will of businesses purchased and which it is deemed inadvisable to carry as an asset .... .01

Deduction is being made for extraordinary hauling charges .. .012
Credited to certificate of indebtedness account and for which after the close of the fiscal year, a certificate of indebtedness bearing interest, and maturing in five years, will be delivered to the members, per hundred pounds .... .15

Total deductions per hundred pounds .... .21

There is nothing in this kind of report to show the total income for milk or the income per hundred pounds. The price reported was not the approximate "price received for all Grade B milk." It did not correctly inform the farmer as to the income received for milk.

These small items of expense in detail led farmers and others to believe that the total of the items represented the total expense. The Federal Trade Commission Report shows that the expense for the year 1924 was not 8 cents but 50 cents per hundred pounds of milk. The Ernst & Ernst audit showed, as above stated, that it was 84 cents in 1937.

The figures published by the League varied at times but producers generally supposed this covered the cost of selling the milk. A statement was given out monthly showing the number of pounds of milk handled, the percentage going into each class and the price for each class. It seemed to be a simple problem to figure the receipts and disbursements for producers from these items in gross figures. But there always seemed to be a portion of the income "not accounted" for. I asked the president of the League to help me strike a balance. Several general explanations were made but none ever supplied a satisfactory answer. When the annual reports were available they revealed the corresponding amount of income unaccounted for. Finally Mr. Sexauer stated that the milk was not all sold at the prices scheduled, but even after the State fixed the price and it was a legal violation to sell for less, the shortage continued.

Ernst & Ernst found only the net returns in the report. But the audit of the books revealed the gross income and an item of $6,055,167.43 deducted as "freight returns and allowances." Pressed for an explanation, the management explained it as "price adjustments which the association was compelled to allow its customers." In other words it was rebates to dealers. Nothing in the record showed where this money went or who got it. This item, together with the other staggering items in the above expense table, accounts for some of my difficulty in making the League's income and outgo balance. This item "freight returns and allowances" appears in all the statements from 1930 to 1937, as well as in previously published reports. It is not a proper deduction from the gross sales. When the milk is dumped from the farmer's can, it passes out of his control. The buyer accepts it. Prices were fixed for the portions of it devoted to different uses. Such were the terms. If rebates were made by the handler, it was his responsibility and not that of the producer.

Missing Vouchers.

Another incident of the audit should be noted. The auditors found a gross item of $385,114.18, reported as paid out on 10,501 vouchers which were missing. The explanation was that they were destroyed to save filing space. This expenditure occurred principally at the time the 1937 Legislature was in session, when it was estimated in the hotels and Capitol halls at Albany that the League officials had spent more than $100,000 to promote the Rogers-Allen bill. The Graves resolution to investigate what was known as the "slush fund" was referred to a committee and stayed there. An appropriation of $50,000 was made in that session for this audit. The officials knew that it was coming and that these vouchers would be asked for.

The League's Operations as a Dealer.

For the year ending March 31, 1937, 14,333,925 cwts. of League members' milk were delivered by the members direct to dealers other than the League. Some 11,246,907 cwts. were received at 106 League plants. About 2,445,000 cwts. of this amount were sold to other dealers, making a total of substantially 16,778,925 cwts. sold by the League and about 8,800,000 cwts. sold by it as a dealer to stores, hotels, and others.

What most dairymen did not know was that the League buys

milk and cream from other dealers and has gone into the merchandising business in eggs, orange drink, tomato juice, chocolate drinks, Bosco, O'Boy and Vizoy, and the reported sales of these items indicated a net loss to League members of $86,000 over a period of eighteen months.

Dealers selling butter, cream and other products to the Dairymen's League from April, 1936 to September, 1937 are recorded as follows:

Name and Location	Item	Cost
Land o'Lakes, Minneapolis, Minn.	Butter	*$322,000.00
New England Dairies, Maine	Cream	*330,812.35
Miller-Richardson Co., Lowville, N.Y.	Cheese	69,053.61
Hawes Dairy Co., Ogdensburg, N.Y.	Butter	*32,897.75
C. W. Burckhalter, New York City	Skim Cond. and Powder	27,432.93
Challenge C. & B. Assn., Los Angeles	Skim Powder	9,200.00
Wisconsin Co-op. Cr. Assn., Elroy, Wis.	Skim Powder	8,243.70
Turtle Lake Creamery Corp., Turtle Lake, Wis.	Cream	7,000.00
Ward Dry Milk Co., St. Paul, Minn.	Skim Cond.	7,487.50
Orrville Milk Cond. Co., Orrville, Ohio	Skim Cond.	6,233.60
Brighton Place Dairy Co., Rochester, N.Y.	Skim	7,547.60
Pioneer Ice Cream Div., Gouverneur, N.Y.	Cream	6,629.00
Borden Farm Prod. of Mich., Detroit	Skim Powder	6,496.89
Goshen Milk Cond. Co., Goshen, Ind.	Skim Cond.	4,049.52
Sparks Dairy Co., Buffalo, N.Y.	Cream	1,250.00

* Approximate.

Other dealers who sold to the League were Woodlawn Farm Dairy Co., Scranton, Pa. (skim condensed), Elroy Co-op. Dairy Assn., Elroy, Wis. (skim powder), Consolidated Badger Co-op. Assn., Shawano, Wis. (skim powder), Fairmont Creamery Co., Omaha, Neb. (cream), Janssen Dairy Co., Hoboken, N.J., and Borden Co. (fluid milk). The volume of this part of the business is indicated in the item of $742,153.47 in the expense schedule on page 291.

Aside from any possible profit or loss, it is not consistent with the best interests of farmers in the New York milkshed to import dairy products from other fields, nor to go into the business of promoting the sale not only of such dairy products but also of milk substitutes in competition with their own milk in their own markets. Farmers themselves never do it. No person or group working for their interests would do it.

The bulk figures in the expense schedule give a farmer information enough to convince him on the face of it that it is costing him too much to sell his milk. It gives him enough information to quit if he were free to do so; but it furnishes no information to enable a group of farmers to sit down together or individually and form a sound opinion on what should, or could be done to make the League an instrument of economic service to farmers. There is no information in the financial statement or elsewhere that would qualify its producers to vote intelligently as members on any business matter pertaining to the management of the organization.

League Shows Heavy Loss.

There is always something lacking in League statements. The auditors found the same embarrassments in the records. Borden's takes 45 per cent of the League's milk. No producer knows what price Borden's has paid for a quart of it for twenty years. The audit does not show. No one knows what price the League gets for the fluid milk sold from its own plants, nor the price and volume sold in the Metropolitan market in competition with other dealers. The record does not show.

The business of selling milk to dealers as an agency is muddled with the business done as a dealer. Both records are mixed in with the merchandising transactions. Any firm doing a private business would keep an exact, separate record of these divisions but trustees handling other peoples' money should be required to do so and to make periodic accountings. Trustees of farm co-operatives may get a degree of satisfaction from their success in avoiding this duty but the record reflects no credit for their sense of moral responsibility.

The analysis of the audit also shows a loss in publishing the *League News*, the official organ, over and above the subscription charged to each producer in an amount of $7,534.61 for the year ending March 31, 1937 and a loss on the sale of milk handled by the League as a dealer to the amount of $11,450,000.

The audit shows operations of the year ending March 31, 1937, in two periods of six months each. In terms of hundredweights, the results can be summarized as follows:

	Sales Price	Expenses	Net Returns
6 mos. 9-30-36	$2.487	$0.667	$1.82
6 mos. 3-31-37	3.01	1.96	1.95
Average for year	2.723	.84	1.88

Further analysis discloses that sales to dealers netted $2.33 per cwt., and that the net return for the 8,800,000 cwts. handled by the League as a dealer was $1.03. Thus, as a dealer, the League lost $1.30 per cwt. on 8,800,000 cwts. of milk or $11,450,000. The other 14 dealers whose books were audited all showed profits.

Attorney General Bennett's Report.

For years League President Sexauer has explained that his failure to stabilize the milk market and pay his producers a reasonable price for milk was due to the cutting of prices by other dealers in the Metropolitan market. He characterized such dealers as chiselers and secured the passage of several laws which he said were intended to stop the practice, but without avail. It was, therefore, rather surprising to read the following apology by himself in the 1938 Report by Attorney General Bennett:

"The Dairymen's League officials explained that a great many of its practices could not be compared with those of other distributors. For example, they indicated that they might be willing to make concessions such as liberal credit, large discounts and money loans, on the surface unbusinesslike, but with the net result of creating a wider and more constant market for the milk of their members. In all business transactions, it was explained, that by reason of its contract obligations to take all of the milk of each of its members, its first consideration must be to find the widest market at a fair price."

This looks like a confession to ward off the penalty for a discovered guilt. Liberal credits and loss of accounts led to taking over bankrupt distributors. "Unbusinesslike discounts" and "money loans" are subtle ways to cut prices and to undermine other distributors. If these practices are virtues on Mr. Sexauer's part, probably creating more business for himself, why was it such a heinous crime for other distributors to imitate his practices?

Mr. Sexauer's complaint has been that such dealers slaughter prices and disturb markets and then recoup themselves for the loss by reducing the price to be paid producers. But it is equally true that Mr. Sexauer has recouped his losses in the self-same way, only to a greater extent. He pays his producers a lower price for their milk than the alleged chiselers pay their producers in the same territory. As to the obligation of taking all of the milk of each of its members, Section 5 of the League contract with each producer indicates that he assumed no such obligation:

"It is further agreed that in case the Association fails to sell the milk of the Producer as herein stated, the producer will in each and every such case manufacture such milk into such products as he may desire and deliver said manufactured product to the Association upon its order for sale, or if directed by the Association, will deliver such milk at such station or manufacturing plant as said Association may direct."

The first consideration of every distributor as well as Mr. Sexauer is to find the "widest market at a fair price" and to protect his market.

No Check on Borden's Reports.

It has been shown that Mr. Sexauer has assured his producers in letters and in other ways answering their complaints that audits were regularly made to determine that dealers made the correct report of the amount of milk used in each of the nine different classifications. On this subject Attorney General Bennett reports (p. 38) that, in answer to a question of what had been done to check the accuracy of the statements of use as provided by various of its customers, "officials of the Dairymen's League testified that monthly audits are conducted." However, Mr. Bennett's examination of these monthly audits discloses that "they are not so much audits as they are copies of sales records furnished to the auditors by the customers of the Dairymen's League. A typical preface to a monthly audit is the following:

"In verifying the disposition of the milk by Borden's Farm Products Division of The Borden Company, we have in accordance with your instructions accepted the reports of their plants pertaining thereto."

The Bennett Report continues: "Apparently no actual check has ever been made by the Dairymen's League to determine any possible misstatement of disposition of milk purchased. *****As far as our inquiry has disclosed, there has never been any actual audit by co-operatives to ascertain the accuracy of the blended price as reported to the producers. By the same token, there is no proof that the statements of use as given to producers are true. No farmers' organization, to our knowledge, has ever determined this all-important fact by actual audit."

Again the Attorney General calls attention to the fact that "The Borden Company is the largest single customer of the Dairymen's League, buying 45 per cent of all of the milk handled by or through the Dairymen's League"; and further that "the sale of 22 per cent of the total dollar volume of milk and milk products at retail and at wholesale by the Dairymen's League, through their own subsidiaries and affiliates, would naturally incline the League to a distributor's point of view."

The Attorney General quotes the Federal Trade Commission as saying that no evidence of dealer domination had been found and that the above remarks by themselves are not an indication to the contrary. "However," he said, "it is only to be assumed that the factors just set forth would have great weight in any determination of policy. The Dairymen's League would have a fuller understanding and possibly more sympathy with distributors' problems than another co-operative might have." The best example of this, he explains, "is the fact that the Dairymen's League voted to reduce the price to producers in December, 1937. Other co-operatives voted to sustain the price."

The Bennett Report accepted statements from Dairymen's League spokesmen and from files which might well give the public erroneous impressions and incorrect information. It includes a table of four years' operation and cost of selling milk (p. 36), which includes 1937. The table shows the cost as 46 cents per cwt. The Ernst & Ernst audit showed the cost at 84 cents per cwt. for that year, the difference being accounted for by the $6,000,000 of rebates paid to dealers, and other similarly disguised items.

The Report, however, shows a desire to find the truth. It contains much reliable and useful information. I have quoted from it freely. It shows a state of affairs that should not be allowed to exist in the dairy business. It is not easy to account for the fact that the State government, in its legislative and executive branches, neglected to use the information to correct the abuses, exploitations and crimes revealed in the Report.

CHAPTER XXXIV

DAIRY LAWS AND THE COURTS

The Co-operative Law.

The fundamental difficulty of the dairy co-operative laws of New York State is that while they are ostensibly intended for the benefit of milk producers they have been revised or written in the interests of middlemen. Some of the authors may have been well intentioned but in practice they have worked out to the disadvantage of the interests that they were supposed to help.

The form of the Co-operative Corporations Law is in itself abhorrent to co-operative principles. It authorizes, as in the case of the Dairymen's League, a centralized corporation of dairy farmers distributed over an area of seven states with one main executive office. The radius exceeds four hundred miles. Members can have no personal touch with it. The law provides for one vote for each member and no proxy votes and then inconsistently provides for proxy votes through delegates.

The present co-operative law was originally written and sponsored by Aaron Sapiro, whose record in co-operation is not what can be suggested for emulation. The statute has been amended and modified since. It has many provisions for the benefit of the corporation and the official management, but no safeguards for the farmer-patron. It was designed particularly for the benefit of leaders of the Dairymen's League Co-operative Association, Inc. and the Borden Company.

In Section 18 of "General Provisions" it prohibits the use of the word "co-operative" by associations not organized under this statute or any of its abbreviations as a part of the name or title of a partnership, corporation or association doing business in the State. Now with the Rogers-Allen Law, dairymen, dissatisfied with the co-operative set-up, are denied the benefits of their own type of co-operation and are confined within a straight jacket if they join the legalized brand. In other words, dealers have captured and appropriated dairy farm co-operation in New York State to their own exclusive use. There are corporations organized under the law that are mere counterfeits of co-operation as it is generally understood and defined.

Section 20 makes it a misdemeanor for anyone to spread false reports, with malice and knowledge, about the finances or management of a co-operative. The offender is subject to a fine ranging from $100 to $1,000 for each offense, and in addition, the co-operative can bring civil suit to collect a $500 penalty. From time to time, the management of the Dairymen's League, which was the original sponsor of this provision, has threatened to enforce Section 20. In November, 1937, it did file a suit against the *Watertown Times* and one of its reporters. The defendant filed an answer but the case was never put on the calendar by the League for trial. It was dismissed finally on motion without trial and without passing on the validity of the law. The decision on dismissal was not appealed. The general law of libel protects co-operatives as others. In any such suit, however, damages would have to be proved. Section 20 and the filing of a lawsuit served to intimidate free speech and free press, but an open trial in court might present embarrassments including an invalidation of the law.

Under Section 21, a co-operative can collect a $100 penalty and obtain an injunction against anyone who is alleged to have induced a producer member to deliver his milk otherwise than to or through the co-operative. The purpose here is to prevent a farmer or a group of farmers from finding a market elsewhere, if for any reason, they feel justified in quitting the Dairymen's League.

At one time two large groups of producers quit two League plants in central New York because of the low price paid by the League. Both groups found a satisfactory outlet in another plant. The court did not grant the injunction asked by the League. The trial was not pressed but the League threatened the buyer on the false claim that the buyer had induced the farmers to quit the

League and the buyer prudently stopped taking the milk. So the League again escaped the need of proving damages; the threat against the buyer served the purpose of the law.

Sections 37, 95 and 120 of the co-operative law provide—each with reference to different types of co-operatives—that under its by-laws a co-operative can assess and collect liquidated damages from any one of its members who refuses to deliver his milk, and further, that such a member can be enjoined from delivering his milk anywhere else. This threat of injunction has always been a favored method of procedure, but not until November, 1937, was the issue squarely placed before the courts of this State. In that case, Sheffield Producers' Association brought a suit against the Jetter Dairy Company, officials of the Dairy Farmers' Union and two individual dairymen, seeking a permanent injunction restraining the dairymen from breaching their alleged membership contract with the plaintiff, restraining Jetter Dairy from taking and buying the milk of these dairymen, and restraining the Union officials from aiding and abetting this transfer of milk from Sheffield to Jetter. Application was also made for a preliminary injunction for the same relief which was denied both by the Supreme Court and by the Appellate Division.

In the lower court's opinion, it was pointed out that a statute giving a co-operative the absolute right to an injunction is of doubtful constitutionality because it deprives the courts of the exercise of their usual discretion in an equity proceeding. The case was ordered to trial before an official referee in Hudson in February, 1938. Testimony by the secretary of Sheffield Producers' Association showed that Sheffield Farms Company had at the same time closed all its fluid plants in St. Lawrence, Franklin, and Clinton counties because it had more fluid milk than its requirements. In consequence, Sheffield Producers' Association expelled the 2000 producers from the twelve large plants in these three counties. This proved that neither the Association nor the Sheffield Farms Company suffered any loss because of the alleged breach of membership contracts. Thereupon the case was abandoned by the plaintiff co-operative, which was so eager to take advantage of the law as a threat, but apparently quite unwilling

DAIRY LAWS AND THE COURTS

to carry through because of the possibility of defeat, which might well have cast a shadow of question on the whole Co-operative Corporations Law.

Section 37 also provides that the member who is alleged to have breached his contract must pay all costs, premiums for bonds, expenses and fees in any action that is brought upon the contract by the co-operative. The by-laws of the Dairymen's League, as authorized by this section, fix liquidated damages at $10 per cow, and if the alleged default continues for more than one month, $3 per cow so long as it continues.

The League contract with its producers is perpetual with the provision that the member may withdraw any year by filing a notice in writing between February 14th and 28th, to take effect on April 1 of the same year. These provisions and dates are embarrassing to the producer. April is the first of the three flush months of milk production. No dealer wants to buy milk at that time, or, if so, he will insist on his own price. No producer would voluntarily select that date to terminate a milk contract.

Sometimes farmers weigh their milk and test it for fat. The plant alone takes the bacteria count. In many cases, the co-operative's weights and tests made at the plant by the co-operative are less than the farmers' weights and tests and its count is variable. The plant manager may return a can or a whole load of milk any day alleging an odor, or temperature or other reasons. He may bar the producer's milk entirely and demand repairs and changes at the farm and barn. He may take his time about making inspections or re-inspections before the milk may again be delivered at the plant. Instances are reported where the farmer was given to understand that a certain place to buy feed would help his case.

In many cases farmers complain that these annoyances follow some complaint or criticism made by the farmer at a meeting or elsewhere. As they express it in confidential letters to me, "if they open their mouths and speak their mind, the weight and test go down and the bacteria count goes up." These reprisals follow also if they neglect to approve any proposal coming from the officers.

Sometimes these conditions become intolerable. The farmer believes that the co-operative breaches its one-sided contract when it makes incorrect weights, tests, or counts and discriminations. He then believes that he is fully justified in selling and delivering his milk to another dealer. Cases are reported to show that when strong and prominent producers do so, the League pays for all milk delivered, and that ends the contract. But in most complaints coming to me the League refuses to pay for milk delivered between the date of last payment and the date shipments stopped, on the grounds that the contract is breached and damages are due it.

The amounts claimed by the producer in each of such cases would probably average $150, plus " certificates of indebtedness" coming due later. After the decision in the Holmes case, the League paid all such claims, but later claims still stand. I have many of them on file. The amount in favor of the League must be considerable, but as far as I know, no report of it has been made.

These liquidated damages are really penalties imposed on farmers by League officers who wrote the law, adjudge the penalty and pronounce judgment. The co-operative suffers no such damages in any instance. The facts in such cases are always in dispute, yet few farmers can afford to pay the expense of carrying a lawsuit through to the Court of Appeals for a month's milk bill. The legal rule is that in case a breach of contract is proved, the injured party is entitled to the amount of damages he actually proves he sustained. Anything more is premeditated robbery.

In the Rogers-Allen Law, the Legislature has further legalized a proxy system of delegate voting. Nothing is more abhorrent to farm co-operation, the essence of which is that each member shall have a full voice in the management and control of his organization. In its 1937 Report to the Congress, the Federal Trade Commission recommended that voting by proxy in co-operative associations be abolished, with this statement:

". . . Each member should have an equal voice in the management of co-operative organizations. . . . The evils of permitting voting by proxy have been noted, and such method of voting by co-operative associations should be eliminated."

Because of this present legalized system of proxy voting, or unit voting as it is more popularly known in the country, the officers and directors of our present large co-operatives can easily, and actually do perpetuate themselves in power. The result is a co-operative organization in name only—in fact, a small minority really acting at its own discretion with no responsibilty to the membership.

Section 19 of the Co-operative Corporations Law enumerates all the powers given to a co-operative association, but it contains no corresponding duties to the membership. We hear professions a-plenty of loyalty to farmers and assurances that the co-operative exists only for its members. But I am not alone in disputing the merit of such claims for the centralized type of associations. In practice in New York their records have come to be regarded as selfish, sinister and ruthless. Judge Cooper found them conspiring to create an illegal monopoly. Investigators have found them deliberately untruthful. Accountants have found them paying millions in rebates and falsifying their books.

The Federal Trade Commission says, in House Document No. 94:

". . . In several instances the management of co-operative organizations were obviously under the influence of distributors and did not adequately protect the interests of producers. . . .

"Dissatisfaction has arisen among certain producers because co-operatives have failed to furnish their members with clear, concise and accurate information as to the disposition of milk and especially as to the method of arriving at the prices paid to producers. . . ."

It must not be forgotten that the classification method of selling milk was a child of the Borden-League alliance.

Again, the Federal Trade Commission says:

"The classified price plan in the absence of audited reports makes it possible for unscrupulous dealers to underpay producers."

Yet, according to the findings in the Bennett investigation, neither the League nor Sheffield Producers, both co-operatives, had ever made an audit of its dealers' books to determine the proper classification of the milk sold to these dealers.

So, there is the Federal Trade Commission stating that classification should not be used unless there is a complete audit of dealers' books, which is not economically possible; and there is also Attorney General Bennett stating that no actual audit has ever been made by the two largest co-operatives in this State. Farmers detest and loathe classification, but even in the face of these official investigations the members of the association are unable to abolish classification.

Section 26 of the co-operative law provides for an annual report of co-operatives and the filing of certain general information in the Department of Agriculture. It does not provide for a profit and loss statement for the information of producers nor any information that would enable farmers to check the accuracy of the returns for their milk or money. In 1932 League officials had this section of the law amended making it a misdemeanor for any person to reveal any information contained in the financial report without the consent of the co-operative, except in obedience to a judicial order.

Just what value to the farmer is a cryptic report in the dark vaults of a government bureau? For eighteen years the Borden Farms Products Company has received all its milk supply from the Dairymen's League Co-operative Association, Inc. and yet there is no League member who has known what price Borden's paid for a single quart of it.

The Milk Control Law.

Attempts to enforce the emergency Milk Control Law (1933-1937) developed some interesting and some absurd situations.

Nebbia Case: A small grocer in Rochester, N. Y., by the name of Nebbia challenged the Milk Control Law by giving away a loaf of bread with the sale of two quarts of milk in violation of the price fixed by the Control Board. Chief Judge Pound, who wrote the decision of the New York Court of Appeals, upheld the law as an emergency measure but held that power to regulate private business can be invoked only "under special circumstances." He justified it in the case before the court in the following sentence: "Price is regulated to protect the farmer from exactions of purchasers against which he cannot protect himself."

Dairy Sealed Case: A Control Board hearing on May 14, 1936 revealed by testimony of its president and its controller that Dairy Sealed, Inc. was a subsidiary of the Borden Company; that it bought its milk from the Dairymen's League; that it handled 50,000 quarts daily in fibre containers during 1935; that the League imported the milk from Pennsylvania at a price that was $88\frac{1}{4}$ cents per hundred pounds less than the New York Control Board price. These officials testified that if they had paid the price fixed by the New York Milk Control Board, it would have cost $300,000 more and if they had used glass bottles it would have cost them $400,000 more.

The testimony showed that the Dairymen's League sold this milk to Dairy Sealed, Inc. and that the Borden subsidiary sold it to the A. & P. Stores, both in violation of the State Law. On June 16, 1936, Dairy Sealed, Inc. and A. & P. Stores were cited to show cause why their licenses should not be revoked. The A. & P. Stores had already admitted the violation as a technical error on their part and paid the difference. There was no evidence that the Borden subsidiary, according to the New York law, repaid anything to its seller nor that the Dairymen's League was threatened with a loss of its license. The information was that the complaint was withdrawn.

Seelig Case: In 1934, G. A. F. Seelig, Inc. was buying milk in Vermont and selling it in New York. The New York Milk Control Board brought suit to compel the Seelig Company to pay Vermont producers the minimum price fixed by the New York Board. The Federal District Court held that the dealer did not have to pay the minimum New York price for milk bought in Vermont, provided the milk was sold in New York in the same cans in which it had been bought in Vermont. As to milk shipped into New York in cans and then bottled and sold in New York, the Court held that the New York Control Board had the right to enforce payment of the minimum prices to Vermont producers. Both sides appealed to the United States Supreme Court. In an opinion by Justice Cardozo handed down on March 4, 1935, it was held that the Seelig Company did not have to pay the New York price in either case, upon the ground that an attempt to regulate prices paid to out-of-state producers was unconstitutional. The Court relied on Section 10, Clause 2 of Article 1 of the Federal Constitution under which States are forbidden without consent of Congress to lay any imposts or duties on imports or exports except what may be absolutely necessary for executing its inspection laws. At the same time, the Court upheld the right of a State to enforce its inspection laws and regulations for the protection of the health and welfare of its citizens by the police powers.

Rosasco Case: This case involved a suit by Elmer Royce, an Oneida County farmer, against John D. Rosasco, who owned three milk plants. The trial was held in Utica before Supreme Court Justice E. N. Smith. Rosasco had reported his sales to the Control Board and paid the blended price to producers. Later the Board re-classified the milk and refused Rosasco a renewal of his license unless he paid producers $55,303.32 alleged to be due them. Mr. Royce sued for $342.21, his share of the amount. It was supposed to be a test case.

The Court found that the re-classification was made on a supposition and presumption by the Control Board without evidence of fact to support it, and that there was nothing in the evidence to challenge the accuracy of Rosasco's verified reports. The Court further found that there was no charge and not a scintilla of evidence of any purpose of conspiracy to cheat or defraud, and that the defendant Rosasco had been made the victim of an arbitrary action. The Court said:

"Here, in revocation proceedings, we have bureaucracy in its starkest form. First, the Milk Control Board (now the Commissioner of Agriculture and Markets) determines what is reasonable notice and publicity for a price-fixing hearing and it prescribes the method of it; i.e., it determines the basis of its jurisdiction. Second, it fixes prices of Official Orders after hearings. Third, it issues licenses to dealers and has the power to revoke them. Fourth, in case of revocation it must give hearings; these presided over by the Milk Control Director, its or his appointee. Fifth, it receives such evidence as it sees fit, and its or his appointee is judge of law and fact. Sixth, it or he has the power to revoke a license. In proceedings to revoke a license there must be no unreasonable arbitrary or tyrannical action, nor may such proceedings be invoked for purposes of punishment, but only in the exercise of police powers to protect the public health, safety or welfare."

The Court granted the motion for a nonsuit and dismissal of the complaint.

In his exhaustive and scholarly opinion in this case, Judge Smith said that the State Legislature has power to provide that no milk shall come within the State for human consumption therein which has not been produced and handled under the same conditions, regulations and inspections as apply to producers within the State; and that the Legislature has the power to "limit the length of the tether of its milk inspection." He reasoned:

"It would be ridiculous to say that to protect the health of the people it would have to inspect all the dairies in the United States, and, unless State health regulations as to fluid milk for human consumption are made equally applicable to out-state milk, the dairy farmers of the State might well complain that they are being denied the equal protection of the laws.

"Legislation effective (a) to secure strict inspection of out-state milk and (b) to use the language of the Court of Appeals in the Nebbia case, 'to protect the farmer from the exactions of purchasers against which he cannot protect himself' would not only be within the legislative power of the State, but would go far to eliminate the necessity of the present complex system of price-fixing. With such legislation, the farmer would be, as he should be, affected by the normal operation of the law of supply and demand."

He quoted Section 6 of Article 1 of the State Constitution:

" 'No person shall . . . be deprived of life, liberty or property without due process of law; nor shall any private property be taken for public use without just compensation.' "

and the Fourteenth Amendment to the Federal Constitution:

" 'No State shall make or enforce any law which shall abridge the privileges and immunities of citizens of the United States; nor shall any State deprive any person of life, liberty or property without due process of law; nor deny to any person within its jurisdiction the equal protection of the laws.' "

and also the Fifth Amendment to the Federal Constitution:

" 'No person shall . . . be deprived of life, liberty or property without due process of law.' "

and stated:

"By and large, the Federal Government has no so-called police powers; subject to prohibitions or limitations, all police powers reside and are inherent in the State governments. The police power is the power in government to protect the public health, public safety, morals, and general welfare of its citizens. This power in State governments is plenary, excepting as limited by express provisions of the fundamental law. But the limitations upon the exercise of this power are: (1) that the exercise of it shall not deprive a person of life, liberty or property without due process of law, and (2) that a State shall not deny to any person within its jurisdiction the equal protection of the laws."

CHAPTER XXXV

DEALERS' SCHEMES

$67,000 In Rebates.

On April 9, 1936, the Pennsylvania Milk Control Board (created under a statute similar to the New York Control Law) cited three dealers of Erie, Pa. to show cause why their licenses as dealers should not be revoked for delinquent payments to the Dairymen's League. The dealers were the Model Dairy Company, Sanitary Farms Dairy, Inc., and Sterling Milk Company. The State auditors testified that the dealers received bills monthly from the Dairymen's League and made monthly remittances of about 80 per cent of the bill and there were never any balances carried forward on the bills from the previous month. The total amount unpaid by the three dealers was exactly $67,716.36 for a period of about 9 months.

The Board held that the amount unpaid on the books was a clever scheme to disguise chiselling prices or rebates and fixed a date to show payments. Then, if not paid, the Board ruled that the licenses would be cancelled. Payments were avoided by another clever ruse. The Dairymen's League reported that it had then accepted notes for the full amount in settlement of the shortage of each dealer. No one, not even the farmers who produced the milk, has authority to inquire if the notes have ever been paid. The Pennsylvania Board formed its own conclusions.

During State Control a State-wide dealer with several plants applied for a renewal of his license. His application was denied on the ground that he had not paid the price set by the Control authorities. This was true. It was also true that the Control price was not being paid by his competitors. He declared himself a "bootlegger in milk" and continued his business without a license. He was not disturbed. Later, his license was renewed.

During the Control period a new co-operative was organized in Oneida County, New York. Its application was denied. After exhausting its efforts to get a license, the co-operative began business without a license and is in business yet. It is my understanding that a license was later issued to this co-operative.

Producer-Dealer Licenses.

Equally absurd situations have arisen under Article 21 of the Agriculture and Markets Law, known as the permanent Milk Control Law. Under this statute a milk producer who sells milk to any buyer except a milk dealer, is defined as a dealer and must be licensed by the Commissioner. To qualify for such a license, the Commissioner must be satisfied as to the character, experience and financial responsibility of the applicant and further that the applicant is not planning to sell milk "in a market already adequately served." That means that a farmer living near a village or a hotel or a summer camp cannot supply such customers with milk even though they may want to buy from him, unless he can show that a dealer is not already in the territory; but every big dealer makes certain that every inch of consumer territory is covered by himself or one of his brethren.

The effect of this law is that few, if any, farmers can get a license to sell their milk at retail to local consumers. They may sell at wholesale to dealers or not at all. The exception is that a farmer may sell not to exceed 10 quarts a day to consumers who come to the farm for it and carry it away in their own containers. Thus, a farmer violates the law if he carries a quart of milk as a gift to his mother across the road from his own premises to her home.

Here are a number of instances involving arbitrary attempts to enforce this licensing law:

In May, 1934, Lewis Selover of Skaneateles, N. Y., was fined $25 on the complaint of a milk inspector for delivering a quart of milk to someone in Auburn without a license. Mr. Selover explained that he had an accredited herd of Guernsey cows, that he had been selling milk for less than the cost of production, and that he did not think he committed a crime in giving away an occasional bottle of milk. He was threatened with another fine if he even gave milk to his 81 year old sick father.

A farmer, who had previously supplied a public institution, put in an estimate and received the award. At a hearing for his license a dealer objected. The farmer's license was refused though the high quality of the milk was admitted, and the farmer's character and responsibility were not and could not be questioned. He was refused a license only because a dealer already licensed desired the trade, put in an objection to the farmer's license, and was sustained.

Another farmer asked his Assemblyman if there was any objection to his retailing milk in his local village. He was told "No," so he went ahead. He provided a suitable building and equipped it at considerable expense to supply village consumers eager to have fresh milk direct from the farm. Then he learned that his Assemblyman, who was in the Legislature when the law was passed, did not know what he voted for. The farmer applied for a license and was refused because local milk dealers objected. The territory was "adequately served" by dealers.

A farmer and his wife had $6,000. They bought a dairy farm from the Federal Land Bank, paid down $4,000 and gave a mortgage with interest payable semi-annually for the balance, retaining $2,000 to buy cattle and equipment, to retail milk in the village to consumers who had been previously canvassed and found willing, in fact eager to have it. The farm couple prudently checked up on their prospects and concluded that they were safe because the returns for the milk would pay the interest on the mortgage and taxes on the farm. No one told them they must have a license before they could sell their milk. Their application was denied. The area was "adequately served." But the Bank insisted on the interest and the collector insisted on the payment of taxes.

During 1938 a producer in a dairy county was denied a license to sell milk in local villages. He proceeded to sell without a license for a few months. In the early part of 1939 he was summoned to appear in the Supreme Court in Albany to show cause why an injunction should not be issued to restrain him from selling milk without a license. After a hearing, Judge Francis Bergan refused to issue the injunction.

From all over the State I get similar reports. Farmers are situated where they could add to their income by delivering their own milk to consumers eager to have it fresh from the farm every morning. The consumers are denied their privilege of buying it and the farmer is refused a license to sell it because in the distribution field the State has granted dealers a milk monopoly.

Until recently there was only one case in the court records of this State dealing with this particular section of the Agriculture and Markets Law. It is known by the name of *Elite Dairy Products* vs. *Ten Eyck*. In that case a license was denied by the Commissioner so the dealer went to court. All that was presented was the determination of the Commissioner denying the license. The Court said that that was not enough, that the Commissioner should submit facts on which he based his conclusion so that the Court could review the facts and then decide whether his determination was correct. Further, the Court expressed grave doubts as to whether a license could be denied upon the ground that the market was already adequately served. The denial of the license was annulled and the case sent back for the Commissioner to prepare and submit the necessary facts so that the determination could then be properly reviewed. I am advised that since that decision was handed down by the Court of Appeals in July, 1936, the Department has never taken any further action in the case, has never submitted the facts required by the Court, and the dealer went right ahead buying and selling his milk without any license whatsoever.

However, in the 1937 Legislature, not quite a year after the decision in this case, that section of the law was amended so as to provide that the Department need merely file a memorandum briefly stating the reasons for the denial or revocation of the license without making any formal findings of fact. Thus the law was amended further to hamstring the producer and the honest dealers because someone was afraid "to go to bat" with the Court of Appeals on whether a man could constitutionally be denied the right to conduct his lawful business.

Under this same Article 21, Section 258-j, no health officer of any county, city or village is permitted to approve any farm producing milk for a plant willing to buy it until the Commissioner is satisfied that the prospective buyer needs more milk and that the existing buyer does not require so much. The real purpose of this provision is to deny the farmer the privilege of changing from one plant to another or to deny him the privilege of selling to any plant at all. In what other field of endeavor can we find an individual prevented by law from selling his products or his services wheresoever and to whomsoever he may choose? Is that liberty?

It would seem from the record that this provision of the law is useful only in denying law-abiding milk producers a privilege to which they are entitled, that producers are justified in disregarding the law and that it should be repealed.

No farmer or friend of dairy farmers ever proposed or approved these laws. They were intended to strip farmers of rights they formerly enjoyed under custom and common law and to give the milk monopoly and its members power to discipline producers individually and make them submissive to dealer domination. Every dairy law enacted in New York State during the past two decades should be entirely rewritten or wholly repealed.

Butterfat and Classification.

Originally the 3 per cent legal standard for butterfat was completely disregarded by dealers in their premium payments to producers, but finally, after many variations, the 3 per cent base was recognized, with 3 cents extra per 100 pounds for each extra point of fat. In 1918, the premium was raised to 4 cents a point. Then in 1926 the Dairymen's League and Borden's adopted a base price for 3.5 per cent milk, retaining the 4 cents premium. The differential was added to the price for every point in excess of 3.5 per cent and deducted when the test showed less than 3.5 per cent. This made a difference of 20 cents a cwt. in the quotation. It was supposed to make the farmer feel better without costing the dealer anything. As a matter of fact, it was a dealers' change and worked out as an advantage only to them.

Milk is not sold to consumers on a guaranteed fat content basis. It is just "milk." If the Board of Health finds milk testing less than 3 per cent fat in a second test, the distributor is cautioned, but 3 per cent is legal and no fault is found. Such milk costs the dealer 20 cents a cwt. less than the quoted price. I have been informed that the average test now seldom runs below 3.25 per cent. This would mean a saving of 10 cents a cwt. to the distributor. Dairy farmers have protested this change in the fat content of the base price, but the change persists and has become a custom. It is another proof of dealer domination. The legal fat test was raised to 3.3 per cent on September 1, 1940 for approved milk.

Milk Returns Compared.

Recently I asked Mr. A. J. Glover, Editor of *Hoard's Dairyman*, if he could tell me the actual returns that Wisconsin dairy farmers receive per one hundred pounds for milk. "That," he replied, "as you know, is the hardest to find and the most uncertain item of information in the dairy business." In a sense I felt personally relieved because I had failed to find the actual net cash paid producers for milk during any particular year or period in New York State. Within the year 1937-38 dairy farmers sent me the actual statements of their buyers showing the weight in pounds and the amount received by check. As far as it goes this is definite and accurate; but, while it is accurate for these individuals, it is not conclusive for other producers or for the whole State.

The Bennett Report printed a table giving the prices per month from May 1, 1922 to December 31, 1937 as the "milk prices paid to members of the Dairymen's League Co-operative Association, Inc., for 100 pounds of milk testing 3.5 per cent fat in the 201-210 mile zone."

On page 43 of the Report is another table under a similar caption to show prices paid members of Sheffield Producers' Cooperative Association for the same months and years.

From these tables I have computed the following averages and comparisons.

YEARLY AVERAGES PER 100 LBS.

Year	Dairymen's League	Sheffield
1921	$2.261	$2.646
1922	2.056	2.50
1923	2.327	2.73
1924	1.988	2.421
1925	2.376	2.666
1926	2.427	2.655
1927	2.585	2.77
1928	2.612	2.762
1929	2.698	2.851
1930	2.318	2.493
1931	1.685	1.891
1932	1.07	1.291
1933	1.219	1.472
1934	1.435	1.711
1935	1.55	1.746
1936	1.666	1.855
1937	1.67	1.903
1938	1.426	1.737
1939	1.563	1.722

The Sheffield figures check correctly and accurately with the returns sent me by Sheffield producers.

The Dairymen's League figures in the monthly statements run from fourteen to twenty-four cents above the actual cash returns paid the farmers who sent me their original return slips. The deductions, which have varied from 20 cents to 5 cents per cwt. for capital funds, were counted in the League returns, but on later returns the capital fund reductions were 6 cents to 5 cents per cwt.

To find the actual cash returns to the farmer, I added the charge for "supplies purchased at plant" to the check and divided the amount by the weight of the milk reduced to hundredweights. This showed the actual net cash return per hundred pounds to League producers who sent me their vouchers.

If these were a fair average of all returns to all producers, the actual cost price paid producers would average about 15 cents less than reported in the League table. The League management makes differentials for plants, cities. regions, etc., the details of which I have never been able to grasp, but since all the milk is pooled it seems evident that any differential advantage given one region or market or producer must of necessity reduce the returns to producers in other areas.

The Classification Swindle.

Classification and blended prices under the Borden-League combination, and now sanctified by law, have made farmers more helpless than ever before. The first "pool" blended price made to producers for 3 per cent Grade B milk for May, 1921 was three cents a quart. The consumer paid fifteen cents a quart. The dealers' spread was 12 cents or more than double the spread in 1915.

The flat price for the same month to non-poolers was 4.9 cents a quart. The dealers' spread was ten cents a quart.

Thus classification increased the dealers' spread 2 cents a quart or ninety-four cents a cwt. for the same grade of milk in the same market. This means that producers were paid ninety-four cents less under the classified system.

For the eight months of 1922 from May to December inclusive, the flat price averaged 59.4 cents per cwt. more to farmers than the "pool" classified-blended price.

For the full year 1923, the flat price averaged 60.4 cents more than the "pool" classified-blended price.

In 1924 Sheffield Farms adopted the classified price plan as a concession to the League officials in an effort to reunite the dairy industry. The attempt failed, but after it adopted classification Sheffield Farms' return to producers approached more to the League price and its spread increased. Prices to all producers dropped with an annual regularity down to 1932 when the price at the farm fell to less than a dollar per hundred pounds.

Increase in Spread.

In my testimony before the Pitcher Committee in 1932, I submitted the following statement to show actual returns to farmers before and after classification:

COMPARATIVE PRICES OF MILK

Year	To Farmer per quart	To Consumer in quart bottles
1915	3.3c	9.0c
1931	3.6c	14.0c

SPREAD FROM PRODUCER TO CONSUMER

1915		5.7c
1931		10.4c

The 1915 farm price was adjusted from a 3 per cent base to a 3.5 per cent base for comparison.

The committee paid no heed to this testimony. On the contrary, the Pitcher Milk Control Law of 1933 gave classification and blended prices the sanction of law. It was confirmed by the Rogers-Allen Law and continues today under the Federal and State Marketing Orders. Classification is a substantial asset to milk dealers and a burdensome liability to milk producers. Dealers want it. Farmers protest it.

Collecting Milk Money.

The collection of milk bills has been an ever-recurring problem for dairy farmers. For many years the farmer who escaped the loss of more than one or two month's bills was said to be in luck. After the State law required dealers to post a bond with the Commissioner of Agriculture, it reduced the losses but never entirely eliminated them. In case of default the bond has rarely been large enough to cover the loss. This is partly due to the fact that the dealer managed to get away with a bond that did not fully cover his credits, and partly due to the fact that farmers did not insist on prompt payments. There is an excuse for both. If the bond is made too high, the smaller dealer is embarrassed for want of capital funds; and at times when other markets are not available farmers naturally take chances and accept the dealer's promises to send the checks; but when credit is extended too long, the dealer finds it more profitable to make an assignment or apply for a receiver. For a considerable number of years conditions in the Federal bankruptcy court in New York City reached the proportions of a national scandal.

It was hoped that the dairy co-operatives would reduce losses on milk bills. Perhaps they have helped. The Dairymen's League has created a fund to be used to cover defaults and bad debts. It makes a deduction out of every bill monthly to create the fund. This spreads the loss over all its producers but it has not stopped the losses. At least three of its losses have been the heaviest producers have ever suffered, but being spread over a large number the burden on each was less. No report of these losses is available.

In and out of bankruptcy courts I have had some interesting experiences. In these courts, with minor exceptions, the experience used to be discouraging. The bankruptcy courts were simply used as a means of cheating producers out of the amounts due them. The bankrupt usually went back into business with capital to do business. One dealer told me he had failed three times but did not make a dollar any time.

I never found it difficult to secure settlements of milk bills due farmers from dealers when I had the complaint within two or three weeks after the date of payment fell due. In such cases the honest dealer felt the necessity of keeping up his general credit and the professional defaulter did not want to default with such limited liabilities. If he had bills due for several months it would be different. It was because of these experiences that I persistently advised producers to insist on prompt returns on the day of payment.

I recall two interesting experiences in the State of Pennsylvania. In one a dealer had given a bond but finally held up payments until the bills amounted to about $4500, as I recall it. The farmers sent the bond to an attorney. After some six months he sent it back with the advice that a signature was lacking, the bonding company held it invalid, and the attorney believed that the indemnity could not be collected. The farmers sent it to *The Rural New-Yorker* about twelve days before Thanksgiving for advice. I went personally to the office of the bonding company in New York and presented the case to an officer of the company, explaining my position as an intermediary without compensation and when successful without publicity. Neither of us referred to the neglected signature. He took a few days to look up the record. Two days before Thanksgiving I called on him again. He was pleasant but wanted more time. I said I wanted the producers to have the check for Thanksgiving and he promised to send the check in the afternoon by messenger so that I could mail it that night. The farmers had the check Thanksgiving morning. They felt happy and so did I.

Some time after, a farmer from the northwestern part of Pennsylvania asked me if a check made out by the Reick, McJunkin Company was good, even if the bank on which it was drawn had failed. I thought it was but to make sure I asked my attorney. Then I wired back "yes." The next day I received some forty odd checks to collect, and as I recall, the amount was about $4,000. The company lawyer in Pittsburgh advised me that in Pennsylvania the failure of the farmers to collect promptly absolved the Reick, McJunkin Company from responsibility.

I took a sleeper that night to Erie, Pa. Through a friend I found a responsible lawyer who advised me that the Pennsylvania courts had held that when the receiver of a check held it for an unreasonable time and the bank failed, the holder of the check was responsible for his neglect. Our checks had been given out Sunday morning and the bank failed on Thursday. The weather was cold and the roads were bad and some checks were in the process of collection through dealers and other banks. I insisted that under the circumstances there was no neglect. Three local attorneys were unwilling to take the case against the Reick, McJunkin Company. We therefore filed suit through Brooks, English, & Quinn, of Erie, for two or three checks. We won a judgment but the company appealed to the higher court and delayed termination for two years. Our judgment was finally affirmed and the producers got their money with interest. The farmers paid the cost of the suit out of the collections.

Following that, I had a case for producers at a cheese factory in St. Lawrence County against a Brooklyn cheese-maker who defaulted owing a month's milk bills to all the producers. He was under bond. We filed suit in St. Lawrence County but the attorneys could not get the summons served. I finally had it sent to me. I asked a friend to serve it and in three hours he came back to report that the service was complete. We got the judgment. Still the dealer fought through technical legal procedures for three or four years and even then the farmers were so wearied with it all they consented in an interview at Albany at the advice of the Department of Agriculture to accept a compromise. I always believed that there was political influence behind the scenes in that case. If I had been in the conference, I should have insisted on every cent including interest; there was no excuse for taking less.

In another case a co-operative plant held a $25,000 bond to secure payment by a Yonkers dealer. Shipments were stopped but the bonding company held up payment. The co-operative asked me to collect it. The company finally asked for $2,500, a 10 per cent discount. The producers were getting anxious for their money and their officers advised me, if I could get a prompt check, that they would allow the discount as the members needed the money.

The next morning I had an interview with the attorney for the bonding company. I insisted on the full amount and told him that the experience had convinced me that we should go to Albany and get a change in the law to protect producers after they had paid the premiums on these security bonds. That seemed to interest him. He wanted to talk it over with the president of the company. After lunch he called me up and said his company would waive the discount and send me the check for the full $25,000. I received it in the mail the next morning.

CHAPTER XXXVI

ESSENTIAL PRINCIPLES VIOLATED

Leadership Selfish.

Membership control, impartial service, correct accounting and equity are fundamentals of successful and permanent farm co-operation.

One selfish purpose alone led the Executive Committee of the Dairymen's League to violate the first of these principles and that error led to the violation of all the other principles, and disqualified them for successful farm leadership. That purpose was their ambition to become milk dealers in New York City at farmers' expense. The first group had this malady and infected the new members as the surviving leaders selected them.

The first evidence of their purpose is in a letter which R. D. Cooper wrote to me in June, 1916. He referred to a meeting of co-operative plant managers called by him to induce them to authorize him to market their milk in the Big City. Incidentally, the project was referred to frequently during the 1916 negotiations. The same purpose revealed itself more definitely and selfishly in 1918 when Mr. Cooper and his associates incorporated the Co-operative Milk Marketing Association to market the milk of League members in the city under their management. When it failed to prosper, they incorporated the ill-famed Country Milk Company, which as previously related, was a gigantic failure. Following this failure and the scrapping of the old League, the same leaders persisted in their ambition by incorporating themselves as a milk dealer corporation of 24 members—the Dairymen's League Co-operative Association, Inc.

The model for this set-up was the holding company and a chain of subsidiaries of the Borden type adapted to a plan of financing through monthly deductions from producers' milk bills.

The immediate effect of incorporating the Dairymen's League Co-operative Association, Inc. was the break-up of producer unity to make Borden's price control possible. Second was the pooling contract forced on producers by the Borden company. This set up the League as a going concern and provided for classification, blended prices and rebates as a compensation to Borden's for the breach of faith to its own producers. The third effect of the new incorporation was the accomplishment of the old ambition of the leaders to become milk dealers at farmers' expense. The only exact information available as to the cost of operation as dealers is the 1937 Ernst & Ernst audit which shows a loss of about $11,000,000.

Self-Perpetuating Leadership.

A self-perpetuating leadership and a centralized autocratic management were essential to a selfish group determined to set itself up as metropolitan milk dealers with the use of farm credit and farm money.

The first indication of the policy of official perpetuation was seen when the old official group under R. D. Cooper's guidance appeared at the first meeting of producers in December, 1916, with a slate fully made up and an avowed purpose to run or ruin the new organization. Again it appeared in that same meeting when one of the group tried and failed to defeat a resolution to organize the producers on a co-operative formula with full and detailed power and control in the producers. The same purpose was indicated in a meeting of the directors in the State of New Jersey the following week when they insolently repudiated the instructions of the producers to prepare a co-operative charter, and voted instead to operate the new body under the old capital stock corporation, which was devised to perpetuate group management by proxy votes. Finally, the purpose was manifested in the cryptic incorporation of the 24 members in 1919, including the pooling contract, classification, the co-operative laws, and in the provisions of its own by-laws and amendments thereof for the election of directors. The official selfishness was magnified in the surrender of the price-making power and in the fruits of farm organization to the Borden Company, subordinating themselves and the association to gain Borden's patronage at any price and to perpetuate their official control of the association.

Autocratic Control.

The determination to establish a centralized autocratic power first manifested itself even prior to the 1916 meeting when the executive group came to me requesting a change in the original co-operative plan that the officials might have power over the membership. Later it was implicit in all the incidents recorded in the above paragraph to perpetuate themselves in control, and in every purpose and practice of their whole record, including such particular incidents as incorporating subsidiaries and affiliates, the fixing of their own salaries and the lack of any check on their expense accounts.

Their despotic purpose was shown in the fact that they devised the whole pooling scheme with Borden's but without consulting farmers, including the contract, the Borden alliance, and in their refusal even to discuss changes in the official plan which was opposed by a major number of the producers. The final demonstration of official despotic power was shown when by their unprotested proxy vote they kept themselves in control of the old company after they had abandoned its service, and appropriated the substantial balance in the old treasury without making an accounting of it to the stockholders.

CHAPTER XXXVII

GOD HELPS THOSE WHO HELP THEMSELVES

Looking Ahead.

In their search for liberty, our forefathers fled from the hardships and tragedies of centralized governments of the Old World. They had had a trial of the tyranny of kingships. In this country they fought for independence and freedom and they were determined to create a new government under the control of the people themselves and pledged to protect the inalienable rights of individuals. They created our American democracy and system of self government. The government is thus directed by public opinion. This directing power begins with the individuals in the homes and spreads to the family, the community, to the county, state and nation, and finally up to the centers of government. Self-government is finally effected by the votes of individuals electing representatives who pledge themselves to act within the framework of the Constitution.

Within a century and a half the United States has developed into the greatest nation in the world. But even an inspired self-government is human. In its devotion to liberty it has allowed groups of individuals and corporations to create trusts and monopolies, boards of trade, and chambers of commerce, national labor organizations and various forms of industrial combines.

The dairy system in New York is a type of the agricultural organization that has thus been created. Each of these groups is a centralized government of its own. Some of them are of the totalitarian type. Most of them are repugnant to democracy. All of them violate the fundamental principles of our American government and, like the dairy type, ride rough-shod over the inalienable liberties of individuals and groups.

The New York dairy system has all the characteristics of the totalitarian state. Though set up and fostered by the State and Federal governments, its subjects have no voice in it. Their individual rights as guaranteed by the Constitution are violated. Dairy farmers who refuse to submit to these tactics are denied access to the market which discriminates against some of its subjects and favors others. With this jurisdiction it holds the power of existence or starvation over the individuals in its realm. Some are forced to work and give up a part of their earnings to a few. This is abhorrent to self-government. Where these repugnant bodies exist, democracy cannot endure.

The American people cannot be won on a frank appeal to support any totalitarian or communistic doctrine but the appeals are not made direct. Many of them are made in the pretense of fealty and friendship. The New York dairy system has not been created by communists to promote their special creed; it was promoted by milk dealers, their subservient sycophants, and subsidized publishers to gain monopoly profits for the dealers and for the aggrandizement of their hirelings. Thus promoting their own purposes, they have violated the American principles and laws, created a dictatorship regime, and promoted the same peril to self-government that Stalin and Hitler and Mussolini have created in Russia, Germany and Italy and that they hope to create in the world.

Under this pretense of friendship, these conspirators intrigue to exploit and rob dairy farmers without the least care that they are creating nurseries for the propagation of Communism.

In these late days of 1940 this subversive influence has become a danger to our freedom and a peril to our liberties. Now that our American people have begun to realize the truth, I am sure that they will assume their responsibilities as citizens of our self-governing nation. They will restore the inherent rights of all and perpetuate the principles for freedom and liberty for which our forefathers fought and died to bequeath to us, and which we owe as an inheritance to our children and to future generations.

The Essence of the Milk Problem.

In essence the milk problem today is just what it was in 1870 and since. Through the seven decades dealers have intrigued and conspired to buy milk from farmers for less than its actual cost of production. Farmers resisted and contended for a fair price. During the whole period the dealers have used sham corporations, in which they held a major control and in which their susceptible country and city stooges have played only minor parts. The corporations were intended to give an appearance of fairness to their deceit. The first two such corporations were ousted as unlawful and fraudulent. A half dozen or more corporations constituting the present system are of the same type and purpose, but more refined and more efficient due to experience and necessity. Originally, contributions to personal political campaigns influenced individual politicians which sometimes led to corrupt legislation detrimental to farmers, but the State government was loyal to producers and consumers. The present system is a counterfeit of farm co-operation devised and promoted under a pretense of benefit to farmers, and as such, has the open support of the State and Federal governments.

Farm resistance reached the fighting point in 1882 and farmers won a victory but were soon again overpowered by strategy and trickery. In 1916 they staged a real fight and won. The sequence to that triumph has been told in preceding pages. It includes what Judge Cooper of the U. S. District Court found on evidence before him to be intrigue, coercion, conspiracy and fraud. It includes the perfidy of farm leaders who conspire with dealers to destroy farm unity and thus destroy farmers' means of attaining justice for themselves. This treachery has kept farmers apart, disputing, confused and helpless for two decades. It was the only way that dealers could recover their lost power to rob milk producers. The only way the dealers can retain that power now, as they well know, is to prevent a reunion of dairy farmers. Hence, they concentrate on that strategy which has netted them handsome results, as can be seen in this twenty-five year record:

	Farmers' Price		Consumers' Price		Dealers' Spread	
	per cwt.	per qt.	per cwt.	per qt.	per cwt.	per qt.
1915	1.62	.033	4.23	.09	2.61	.056
1921	2.65	.056	7.05	.15	4.40	.094
1935-36	1.53	.032	6.11	.13	4.58	.098
1937-40	1.48	.032	6.35	.136	4.86	.103

Classification and political control have thus reduced farm prices and increased consumer costs both of which have doubled dealers' profits, which is just what they were designed to do. Under the present system farmers are therefore denied justice, liberty and property rights guaranteed them by the Constitution. Their only hope is to regain these inherent rights.

A Simple Milk Program.

What the dairymen need is a simple and economic plan for the sale of their milk. For this they need an organization with two separate and distinct parts:

1. The Fluid Milk Division, and
2. The Manufacturing Milk Division.

The Fluid Milk Division will set a price for fluid milk and cream. It will sell all that the dealers will take and pay for at that price. Under present conditions the farm price would be from five to six cents a quart or an average for the year of about $2.75 a cwt. for 3.5 per cent milk.

The cost of distribution through stores to consumers is estimated to be:

Cost	Cents
Cost of milk	.0585
Country plant	.0025
Freight	.0050
Pasteurizing and bottling	.0060
Cartage to store	.0050
Store profit	.0100
Overhead	.0130
Consumer cost	.1000

For milk sold to hotels, hospitals, drug stores, bakeries and other institutions, there will be no store profit and besides a saving on bulk deliveries. This will increase the farmers' income to about two cents a quart or 94 cents a cwt. on all such bulk sales.

This estimate will save consumers five cents a quart. The increased consumption will add to farmers' returns. These estimates are confirmed by men in the city trade.

Consumers who can afford door delivery and insist on it will be willing to pay for the extra service, but it is expected that all stores will deliver milk to consumers. Most of them do now.

If all the fluid milk is not sold, the Fluid Milk Division will make return to the producer for his pro rata share of fluid sales, and turn the unsold milk over to the Manufacturing Milk Division, unless the producer elects to use his unsold portion at home. The Manufacturing Milk Division will receive this overflow of fluid milk, as well as milk from producers who do not wish to qualify for the fluid trade. It will sell this milk to processors or manufacture it into by-products. It will have the whole nation for a market. In a radius of 400 miles it has the best consuming dairy food market in the world. Right here we can develop a wholesale and retail trade of milk products that will tax our producing forces to supply.

Dealers' Plans Have Failed.

Governor Lehman, Commissioner Noyes and some legislative leaders admit that dairy farmers have a right to collectively run their own business and to negotiate the price and sale of their milk, "but," they say, a farm organization "will not work" because "farmers cannot agree," and therefore the dealers, their stooges and politicians must be clothed with legal authority to create a system, to do what farmers cannot do for themselves. Hence, we have the present system, designed by our milk magnates and sanctified by State and Federal authorities and law. Does their system work? It does not. Do its proponents agree? They do not. Their plans have failed dismally. Secretary of Agriculture Claude R. Wickard and Market Administrator N. J. Cladakis proposed eight amendments to the Federal Order.

Seven of them proposed small deductions from the spoils taken out of producers' returns estimated to be $5,000,000 a year. One amendment gave the Administrator authority to find out where this spoils money goes and who gets it. The disagreement became vociferous and loud. The stooges rushed into the Federal Court in defense of their spoils. They asked for a temporary injunction to restrain the Secretary of Agriculture from holding the new referendum and also from suspending the Order pending trial of the dispute on its merits. The court granted the temporary injunction and the end is not yet.

Recently a boy was arrested in one of our states and sent to jail for three years for stealing an apple. For seventy years milk dealers and their stooges have been, in effect, every day stealing value out of millions of quarts of milk. They have exacted high prices from consumers. They have beggared farmers and starved city children.

Some years ago I paid $6,000 for an automobile. In 1940 I bought a better car for less than $900. That is an example of what a free producing industry can do for the people. If the government set up a system of dealers to distribute automobiles, to fix prices for the dealers to pay manufacturers and left the dealers free to charge the users what they liked, the automobile industry would have the same type of system that now cripples the milk industry.

An Appeal for Unity.

In conclusion, I want to make an appeal to dairy farmers. I started to study the dairy business and to work for justice to all in it at the age of 14 years and have continued it since. In the natural order of things at the age of four score and four years, I cannot expect to continue the work indefinitely. In this book I have tried to leave you the benefit, if any, of my study and experience. After the 1916 fight I thought the forty-year battle was won. It was won but we did not hold our gains. I do not consider the present status as final. We won then through unity. You can, and I believe you will, win again through the unity of dairymen who have the will to fight for justice.

Do not think that others can or will do the job for you. You cannot win permanently unless you do it yourselves. That does not mean that each one of you must take a hand in the details all the time. It does mean, however, that most of you must know what is being done, why it is done and have a voice in directing what is to be done. That is what I have meant when I have said "You must do it yourselves." You will need a definite organization and you must help direct it. I think I know farmers and farmers' sentiments and the farm way of life. It was all bred into me in a farm home and through lifelong contacts. I know you have the ability to conduct this dairy distribution business. It does not require as much ability as the operation of your farms. Your sons and daughters climb to the top rounds of the ladder in business, banking, manufacturing and in the professions. They will do the same in dairy distribution, and if you take the initiative now, you will blaze the trail for them.

Together in 1916 we did a pretty good job. I have written this book that you may know what has been the cause of failure since. Borden's spokesmen said then, "The end is not yet. If farmers want organization we will give them enough of it." To that boast I would like to hear dairy farmers say, "In twenty-four years we have caught on to your tricks, your intrigue and your conspiracies. It has been an expensive lesson, but we owe you no ill will. We have no threats to hurl back at you and your kind. You denied us justice and we will submit to injustice no longer. We have a duty in the future to ourselves, to our children and to God. We do not hate you. We do hate fraud and injustice. We have not fought you. We have fought the evils and injustice you have perpetuated. God is the spirit of justice. When we fight for truth we are His partners in the fight for justice, and from now on we will do our part in the partnership."

In the past we have not done our full duty, and have not done what we attempted in the best way. We lacked information and resources. We hesitated to assume responsibility and trusted to others to do for us what we should have done for ourselves. Now we have the record of the past and the information of the present to guide us. The time is right. If we face the task with charity, unity and justice as our guides, we can in sixty days restore prosperity to our dairy industry and make it permanent.

Prosperity in the dairy business awaits the united action of farmers. To them I say—Adopt a true co-operative plan and I pledge myself to work in the ranks with you to the limit of my strength, my ability and my resources.

INDEX

Abbott's Dairies, 47
Adair, Hugh, 108, 195
Addison Co. Co-op. Dairy Co., 146
Agriculture and Markets Law, N.Y. State, 312
Agricultural Council, 214
Agriculture, N.Y. State Dept. of, 55, 126, 214, 256
Agricultural Society, N.Y. State, 55, 56
Allen, Howard N., 228
Anderson, John, 154
Anglo-Swiss Condensery Co., 15
Arfman, John, 108, 109, 150, 151, 152, 159
Arnold, Thurman, 263
Ayres, W. A., 285, 286

Babcock, Dr. S. M., 24
Babcock Tester, 23, 41, 42
Bailey, Oscar, 108
Baldwin, Charles H., 21, 207, 209, 210, 214, 231, 232, 233
Bargaining Agency, Niagara Frontier, 256
Bargaining Agency, Distributors, 225, 231, 239
Bargaining Agency, Producers, 225, 228, 238, 239, 264, 267, 268
Bargaining Agency Prices, 235, 238
Barry, Patrick, 56
Battle, George Gordon, 156
Beach, Fred H., 9
Beakes, Charles H. C., 7, 9
Beardsley, J. D., 108
Bennett, John J., 235
Bennett Report, 211, 212, 229, 230, 236, 274, 278, 296, 297, 298, 316
Bergan, Judge Francis, 257
Big-3, 217, 228, 229, 234, 235, 266, 270
Borden Co., Edward McGuire vs., 192
Borden's, 35, 36, 40, 46, 47, 87, 88, 92, 97, 113, 114, 117, 121, 132, 133, 134, 135, 157, 184, 186, 188, 191, 194, 196, 200, 208, 212, 228, 231, 236, 261, 274, 275, 276, 277, 282, 283, 284, 285, 289, 295, 298, 315 324, 325
Borden's, 1936 Profits, 289
Borden-League Alliance, 186, 188, 189, 191, 195, 196, 201, 221, 239, 274, 275, 277, 286, 318, 325
Borden-League Alliance, Producers Complain, 282
Borden's Condensed Milk Co., 16, 34, 38
Borden, Gail, 16, 23
Borden Wieland, Inc., 261
Boshart, Fred, 56, 108, 195
Brill, Frank, 108
Brill, Jacob S., 83, 96, 97, 98, 99, 109
Briggs, Dr. Herman M., 157
Brockway, Albert L., 160
Brown, Commissioner, 17
Brown, Elon, 72, 73, 112, 122, 125, 126, 127, 128, 129
Brown, Josiah K., 21, 22
Bullville, Farmers' Loss, 145, 146
Burkett, Charles W., 127, 128
Burritt, M. C., 100
Butter, Out-of-State, 294
Byrne, William, 206

Chicago Milk Indictments, 261, 262
City Plant, Plan for, 140
Cladakis, N. J., 331
Clark, John D., 190
Classification, May 1921, 178
Classified Price Plan, 179, 180
Clover Farms, 198, 202
Conklin, George, 7
Consolidated Milk Exchange, Ltd., 9, 11, 34, 35, 36, 37, 38, 40
Cook, Herbert E., 108, 189
Cooper, Judge Frank, 242, 247, 250, 251, 252, 253, 254, 255, 256, 258, 305, 328
Cooper, R. D., 76, 77, 82, 87, 93, 94, 98, 102, 108, 110, 115, 127, 134, 135, 136, 140, 141, 143, 144, 145, 146, 147, 152,

160, 163, 164, 172, 181, 182, 193, 323, 324
Co-operation, Formula of, 105, 329, 330
Co-operation, Prize Plan, 155
Co-operation, Recommendations on, 287
Co-operative Corporations Law, 221, 300, 301, 302, 303, 304, 305, 306
Co-operative Exemptions, 212
Co-operative Milk Marketing Assn., 134, 140, 141, 146, 161, 323
Copeland, Royal S., 143, 157
Country Milk Company, 140, 141, 143, 144, 145, 146, 147, 151, 161, 323
Craig, J. Leslie, 108, 195
Cream, Out-of-State, 294
Cross, James, 203
Culver, Harry W., 70, 108

Dairy Farmers' Union, 218, 235, 238, 239, 264, 302
Dairymen's League Co-op. Assn., 47, 161, 178, 184, 186, 191, 194, 196, 200, 201, 204, 206, 209, 212, 221, 228, 231, 249, 269, 274, 275, 276, 277, 278, 279, 280, 282, 284, 285, 286, 289, 292, 293, 295, 298, 300, 305, 307, 311, 315, 316, 317, 324
Dairymen's League Co-op. Assn., Inc., Charter Filed, 176
Dairymen's League Co-op. Assn., Contract with Producers, 297, 303, 304
Dairymen's League Co-op. Assn., Control of Country Plants, 201, 202
Dairymen's League Co-op. Assn., Loss as a Dealer, 293, 295, 296
Dairymen's League Co-op. Assn., Missing Vouchers, 293
Dairymen's League Co-op. Assn., Rebates to Dealers, 292, 293
Dairymen's League Co-op., Inc., Subsidiaries of, 290
Dairymen's League Co-op. Assn., 24 Members; 24 Directors, 177, 178, 181, 183, 186, 187, 197, 323, 324
Dairymen's League, Inc., 69, 75, 77, 83, 87, 89, 127, 159, 161, 178, 184, 185, 186
Dairymen's League, Inc., Abandonment of, 176, 180
Dairymen's League, Inc., Meetings, December, 1918, 148; January, 1919, 152; December, 1919, 163; December, 1920, 169; December, 1921, 184
Dairymen's League, Inc., 1919 Income and Expense, 165
Dairy Sealed, Inc., Case, 307
Davidson, M. W., 109
Day, Jonathan C., 157
Dealer Corporations, Rule by, 273
De Laval Separator, 23
Denniston, Augustus, 55
Donnelly Act, 225
Durland, Jesse, 7

Eastman, E. R., 181
Elite Dairy Products vs. Ten Eyck, 314
Ely, Alfred, 9, 11
Empire Dairy Company, 201
Ernst and Ernst, Audit, 193, 274, 277, 278, 286, 288, 289, 292, 299, 324

Fair Price Milk Committee, 157
Farm Bureau, 32, 100
Farmers' Alliance, 31
Farms and Markets Bill, 128, 129
Farms and Markets Council, 130, 131
Farm Price Index, 28, 30, 33
Federal Food Administrator, 134, 147
Federal Milk Committee, 135, 136, 143, 147, 158
Federal Land Bank, 59
Federal Order, 240, 250, 251
Federal-State Orders, 240, 241, 243, 244, 245, 246, 247, 248, 249, 268, 271
Federal State Orders, Referendum, 241, 242, 243
Federal State Orders, Unpriced Milk, 246, 247
Federal Trade Commission, 193, 208, 211, 276, 277, 282, 283, 284, 286, 287, 289, 306
Federal Trade Commission Report (House Doc. No. 94), 286, 287, 304, 305
Federal Trade Commission Report (House Doc. No. 95), 276, 277, 278, 282, 283, 285, 286, 292
Federal Trade Commission Report (House Doc. No. 152), 284
Fee, Kenneth, 207
Finley, Dr. John H., 157
Five States Milk Producers Assn., 11
Flanders, George L., 21
Foods and Markets Committee, 119, 122

INDEX

Foods and Markets, Division of, 156
Foods and Markets, N.Y. State Dept.
  of, 62, 64, 65, 67, 69, 84, 85, 91, 120, 126, 128, 140, 160
Food Supply Committee, Mayor Mitchell's, 119
Forristall, E. H., 100
Freight Rates, 51, 203
Fuller, Bradley, 159, 160, 161, 165, 181, 183

General Ice Cream Corporation, 261
Gerow, John Y., 79, 80
Gilmore, J. D., 17
Glover, A. J., 316
Glynn, George, 120, 131
Glynn, Martin, 59, 62, 63, 65, 66, 156
Graves, Rhoda Fox, 288, 293

Halliday, Clark, 108, 195
Harbison, Thomas B., 9
Harden, L. M., 108
Harmon, E. M., 245
Hartshorn, A. A., 80, 81, 82
Health, N.Y. City Board of, 19, 20, 22, 23, 43, 47, 93, 134
Health Permits, 315
Hearst, Mrs. Wm. R., 157
N.Y. Herald Tribune—Reprints from Price Cutting (Mar. 18, 23, 1939), 269, 270
Herrick, Myron T., 59
Hoard's Dairymen, 200
Holmes Case, 173, 174, 175, 198, 199, 304
Hoover, Herbert, 134
Horton, Loton, 88, 158, 195, 196
Hovey, Vernon F., 261
Howell, Amzi, 5
Huson, Calvin J., 21, 56
Hylan, Mayor John, 143

Interstate Commerce Commission, 51, 53
Interstate Milk Prod. Assn., 47, 285
Ithaca-Cornell Bill, 222, 223, 224, 228

Jetter Dairy Co., 302
Jetter-Rock Royal Case, 242, 250, 251, 252, 253, 254, 255, 256, 258, 259, 260, 261
Jones, Homer, 195

Jones, R. V., 261
Jordon, J. V., 9

Kelly, Mrs. Edward, 282
Kershaw, H. J., 79, 80, 82, 160, 161
Keystone Dairy Co., 210
Kirkland, Leigh G., 160
Kohns, Lee, 157
Koppleman, Herman P., 276
Krause, W., 282

Lacy, F. H., 68, 100
Laemmle, Joseph, 7
LaGuardia, Fiorello H., 262, 265, 266
Laidlaw, Earl, 160
Lane, Milton, 150, 151, 152, 159
Latimer, J. C., 13
Lederle, Dr. E. J., 48
Lehman Brothers, 224
Lehman, Herbert A., 206, 207, 208, 214, 221, 275, 330
Lewis, Madison H., 261
Licenses to Dealers, 308, 311, 314
Licenses to Producers, 312, 313, 314
Loeb, Sophie Irene, 157
Loomis, A. G., 12
Lowell, Seth J., 119
Lynn, Preston P., 157
Lyon, H. H., 80

McBride, John A., 9
McCrossin, Edward J., 146
McElroy, Francis, 221
McElroy-Young Bill, 221
McGuire, Edward, 192
McInnerney, Thomas H., 261
Machold, H. E., 72
Mack, John, 92
Manley, Henry, 193, 215, 216
Manning, Albert, 75, 93, 98, 108, 109, 160
Marcussen, William H., 269
Mather, Fred E., 195
Mayor Mitchell's Food Supply Com., 119
Medbury, George W., 282
Membership Corporations Law, 159
Milbank, Albert, 87, 192
Milk, Adulteration, 20, 21, 22, 23, 42
Milk—Advertised Brands, 205, 234
Milk—Ban on Loose, 205
Milk Charter Bill, 220, 221, 222, 223, 226, 227, 231, 288

Milk, Chicago Indictments, 261, 262
Milk, Classification, 206, 223, 315, 316, 318, 319
Milk Committee, 47, 219, 220, 288
Milk Committee's Plan, 219, 220, 221, 329, 330
Milk Conference Board, 115, 117, 150
Milk, Connecticut Dealers, 284
Milk Consumer's Protective Committee, 240
Milk Consumption, 234
Milk Control Board, N.Y. State, 203, 204, 206, 207, 210, 212, 213, 215, 216, 307, 308
Milk Control Board, Pennsylvania, 311
Milk Control Law, N.Y. State, 203, 204, 205, 207, 209, 222, 225, 306, 307, 308, 312, 319
Milk Control Prices—N.Y. State, 210, 213
Milk—Dairymen's League Cost in Wholesaling, 279
Milk Dealers Protective Assn., 38, 39
Milk—Dealers' Spread, 237, 268, 281, 318
Milk—Differentials, 283
Milk Exchange, Ltd., 7, 8, 37, 40, 273
Milk Grades, 43, 44, 45, 133, 147, 245, 281
Milk—Grade A Abolished, 281
Milk Investigation (1940), 262, 263
Milk Meetings, Utica (Sept. 1916), 84; Utica (Dec. 1916), 108
Milk, Out-of-State, 307
Milk, Philadelphia dealers, 284, 285
Milk Plants Eliminated, 283, 284
Milk Producers Union, 8
Milk Strike—1919, 150, 152, 153
Milk Strike—1933, 207
Milk Strike—1939, 264, 265, 266
Milk, Swill, 19, 24, 25
Milk Tolls on Farmers, 271, 272
Milk, Vitamin D, 281
Miller, Allen D., 240
Miller, Burt, 209
Miller, J. D., 7
Miller, John D., 108, 114, 115, 116, 117, 127, 135, 145, 147, 164, 170, 183, 184, 189, 190
Miller, Nathan, 181
Milliman, Thomas, 100
Modern Dairy Company, 134
Muller Dairies, 209, 210

National Dairy Products Corp., 196, 218, 224, 261, 276, 285
Nebbia Case, 306
Nestle Company, 170
New York Dairymen's Assn., 189
New York World, Reprint from, 113
Non-pooling Co-op. Assn., Inc., 184
Norberg, P. H., 154, 155
North, Dr. Charles E., 166
Noyes, Holton V., 21, 241, 251, 256, 266, 275, 278, 288, 331
Nunan-Allen Law, 257

Olmstead, G. O., 7
O'Malley, E. R. (Atty.-Gen.), 35, 37, 39, 73, 85
Orange County Milk Assn., 2
Orange County Milk War, 6, 40

Parran, Dr. Thomas, Jr., 207
Pearson, Raymond A., 21, 55, 56
Pembleton, John G., 160
Perkins, George W., 119, 120, 122, 123, 125, 126, 128, 129, 130
Pitcher Committee, 1932-33 Investigation, 203, 204, 208, 280, 318
Pitcher, Perley, 203
Plymouth (Wisc.) Cheese Exchange, 211
Pooling Contract, 167, 168, 170, 172, 175, 183, 186, 196, 199, 288
Porter, Eugene H., 156
Porter, George C., 154
Poughkeepsie Courier, Reprint from, 67, 68
Pratt, Willard R., 173, 185
Provisional Contract, 1932, 196, 197
Pure Milk Company, 14
Pyrke, Berne A., 21, 56

Rathbun, H. H., 282
Reick, McJunkin Co., 321
Rogers-Allen Bill, 224
Rogers-Allen Bill, Slush Fund, 293
Rogers-Allen Law, 225, 226, 227, 228, 229, 232, 237, 238, 257, 267, 271, 300, 304, 319
Rogers, George F., 212, 228
Rosasco Case, 308, 309, 310
Rural New Yorker, Reprints from Washingtonville (N. Y.), Protest Resolution, 46; On 1916 Fight, 70, 71; Borden-League Plan, 136, 137, 138;

INDEX

Farm Co-operation Advised, 153; Farm Letters, 162, 163

Sanford, M. L., 9
Sanford, P. E., 7
Sapiro, Aaron, 300
Schraub, Frederick, 21
Seelig Case, 307
Selover, Lewis, 312
Sessions, Fred, 56, 108
Seven States Pact, 214, 227
Sexauer, Fred H., 144, 238, 239, 274, 282, 283, 284, 288, 292, 296, 297
Sheffield Farms Company, 36, 38, 47, 92, 135, 158, 190, 195, 196, 197, 204, 210, 212, 228, 231, 235, 236, 238, 267, 275, 285, 289, 302, 318
Sheffield Farms Company, 1936 Profits, 289
Sheffield Producers Co-op. Assn., 210, 285, 302, 305, 316, 317
Sherman, Frank, 76
Sims, Clifford S., 119
Sisson, Fred W., 276
Sisson, George, 56
Siver, Robert W., 160
Slaughter, George, 7, 9
Slocum, George W., 171, 172, 176, 183, 200
Smith, Alfred E., 156
Smith, Justice Edward N., 229, 308, 309, 310
Smith, Frank M., 160, 161
Smith, Thomas O., 7
Spencer, Dr. Leland, 203, 208, 261
State Dairy Commission, 21
State Food Commission, 130
State Land Bank, 58
State Order, 256, 257, 258
State Standing Committee, 57, 60
Stevens, Robert F., 7
Stoddard Mfg. Company, 23
Straus, Nathan, 23
Sweet, H. B., 185
Sweet, Thaddeus, 63, 64, 66

Tabor, Charles F. (Atty.-Gen.), 8
Tayntor, John W., 7
Ten Eyck, Peter G., 21, 190, 210, 214, 227
T. B. Test, 23
Thompson, F. H., 70, 76, 90, 93, 94, 97, 98, 109, 160

Thompson, William Boyce, 118
Todd, J. H. L., 282
Tone, R. R., 7
Towner, Bill, 112, 116, 127
Towner, Senator, 112, 113
Tuthill, T. J., 7
Tuttle, Ezra, 56
Tyler, George M., 160, 161

U.S. Supreme Court, 258, 259, 260, 261
Unity Dairymen's Co-op. Assn., 190
Utica Press—Reprints from R. D. Cooper letter, 193
Utts, V. W., 282

Van Son, N. A., 89, 141, 142, 143, 144, 147
Vitamin D—Profits in, 281

Wallace, Henry A., 251
Ward, George W., 75, 119, 121, 122, 123, 124, 125, 126
Warren, George, 97
Washington Square Pump, 22
Watertown Times, 301
Weiant, C. A., 136, 168
Wells, George E., 11
Wells, J. E., 9
White, Charles, 56
Whitman, Charles, 73, 119, 122, 123, 126, 128, 129, 130, 131
Whitney, Dr. Caroline, 240
Wickard, Claude R., 330
Wicks Bill, 123, 124, 126
Wicks, Charles W., 72, 73, 75, 119, 121, 123, 126
Wicks Committee, 42, 49, 72, 119, 121
Wierck, John P., 9
Wieting, Charles A., 21
Wilson, Charles S., 21, 156
Winchester, Mrs. John, 283
Wisconsin Co-operative Plan, 182
Witt, William C. A., 9
Woodbury, W. W. (Atty. Gen.), 90, 91, 95, 96
Woodward, Judge Charles E., 261
World War, Prices, 147
Wright, Archie, 264, 265
Wright, W. A., 7, 9

Young, Fred, 221

Zimmer, M. L., 282

INDEX OF TABLES

Milk Prices, 1870-95, 17
Borden and Exchange Prices, 1909, 36
Fat and Bacteria Premiums, 44
Milk Yield per Dairy (1902-16), 50
Freight Rates, 52, 53, 54
Consumer Prices (1934-1937), 211
Milk Dealers' Spread, (1912-1937), 281
Dairymen's League 1937 expenses, 290, 291
Form of Dairymen's League monthly expense account, 291, 292
Dairymen's League and Sheffield prices (1921-1939), 317
Milk Price Comparison (1915-31), 319
Farmers' Price, Consumers' Price, Dealers' Spread (1915-1940), 329
Distribution Costs, 329
Milk Price Comparison (1936-7-8), 236
Analysis of Dealer Spread (1937-8), 237, 238
September, 1936-37-38, Milk Prices, 248
Analysis of Dealer Spread (1938-40), 268
Cost of Wholesaling Dairymen's League Milk, 279
Cost of Farm Supplies in Terms of Milk, 280

Made in United States
North Haven, CT
02 December 2023